# Intelligence Sharing Between Asymmetrical Allies: The US, Uganda, Sudan, and South Sudan Against the LRA

**Malual Ayom Dor, BA; MA; MRes; PhD**

*A Note from the Publisher*

The publisher wishes to acknowledge and thank Dr Douglas H. Johnson for his invaluable help and support for Africa World Books and its mission of preserving and promoting African cultural and literary traditions and history. Dr Johnson and fellow historians have been instrumental in ensuring that African people remain connected to their past and their identity. Africa World Books is proud to carry on this mission.

Cover design, typesetting and layout : Africa World Books

# Dedication

For the many victims of the LRA.

# Table of Contents

## Chapter 3: The Peaceful Approach

## Chapter 4: International Relations

## Chapter 5: The Military Approach

## Chapter 6: War by Proxy

## Chapter 7: The Importance of Intelligence Sharing

**Chapter 11:**
**The ICC and the Indictment of Joseph Kony**

**Chapter 12: LRA Defeated but Still at Large**

# Abstract

This book discusses the importance of intelligence sharing between asymmetrical allies in the fight against the Lord's Resistance Army (LRA). The book is unique in that it documents and examines intelligence sharing between the US, Uganda, Sudan, and South Sudan and is written by an insider, who served in a senior position in the SPLA when intelligence sharing was taking place among these countries. This book is useful to scholars and practitioners in the fields of intelligence, defence, security, and strategic studies and can be used in universities and military and security academies. The book underlines the importance and contribution of intelligence sharing against the LRA, a merciless Ugandan rebel movement that caused havoc in Uganda, as well as the neighbouring countries of South Sudan, the Democratic Republic of Congo (DRC), and the Central African Republic (CAR) since it began its protracted war against the government of President Museveni in 1986.

The book aims to demonstrate the ways in which the LRA's capacity was drastically reduced by the intelligence sharing that took place between the US, Uganda, and South Sudan. This intelligence sharing transformed the LRA into the theoretical organisation that it is today. What is clear, however, is that asymmetrical partnerships are successful where the stronger partner remains committed to the partnership. Indeed, when the US announced on 26 April 2017 that its special forces would cease all operations against the LRA, the Ugandan People's Defence Forces (UPDF) followed suit and announced that it would immediately withdraw due to logistical reasons. Subsequently, South Sudan announced that, in the absence of US and Ugandan forces, it would not continue with LRA operations, thus clearly indicating the gap created where the stronger partner leaves an asymmetrical partnership. Crucially, this book contends that the chance of the LRA resurfacing cannot be ruled out in the absence of effective intelligence sharing between the asymmetrical partners.

# About the Author

Lieutenant-General Dr. Malual Ayom Dor is a founding member of the SPLM/A. He has held various senior positions in the SPLA, including those of Deputy Chief of General Staff for Administration; Assistant Chief of Defence Forces for Operations, Training and Intelligence; Inspector General of the SPLA; and Director for Military Training and Colleges. He is currently an Assistant Professor of Security and Strategic Studies at the Institute for Peace, Development and Security Studies (IPDSS) at the University of Juba, South Sudan. The title of his first book is *Learning Through Negotiation: The Role of the SPLM/A in Ending Sudan's Civil War, 1983-2005*. He holds a PhD and MRes in War Studies from King's College London, University of London, United Kingdom; an MA in International Relations and Diplomatic Studies from Makerere University, Uganda; and a BA (*Cumlaude*) in International Relations from the United States International University, Kenya. He attended Oxford University Summer School on International Human Rights Law, Peking LSE Summer School on

Chinese Politics. He was awarded a Chevening Fellowship in 2006 which allowed him study Democracy, Rule of Law and Human Rights at the University of Birmingham. He was among the first batch of SPLM cadres to attend Diplomatic Training at the Netherland Institute of International Relations 'Clingendael' in 2003. He has also attended International Humanitarian Law Training at the United Nations International Human Humanitarian Institute San Remo Italy in 2004. He is currently pursuing an LLB degree from the University of Nairobi, Kenya. His areas of interests include strategic and security studies, diplomacy, foreign policy, international relations and international law. In 2019, he was awarded five medals of service by the Commander in Chief of the South Sudan People's Defence Forces, namely medals for valour, independence, long-distinguished military service, resistance and revolution.

# Acknowledgements

I am very grateful to all those who agreed to be interviewed for the content of this book. Without their contributions, the publication of this book would not have been possible. In particular, I want to mention former Uganda Chief of Defence Forces General Aronda Nyaikirima and former Uganda Chief of Military Intelligence Brigadier James Mugira. It is unfortunate that General Aronda is not alive to read this book.

I would like to acknowledge my colleagues in the SPLA, particularly General James Hoth Mai, the former SPLA Chief of General Staff, and Lieutenant-General Wilson Deng Kuoirot, the former SPLA Deputy Chief of General Staff for Operations, who represented SPLA in the LRA Juba peace talks.

Special thanks go to AWB Book Publishers for publishing this book.

# Definitions of Acronyms Used

Amb. – Ambassador

A/CDF – Assistant Chief of Defence forces

ADF – Alliance for Democratic Forces

AU – African Union

Brig. – Brigadier

Capt. – Captain

CAR – Central Africa Republic

CDC – Centre For Disease Control

CDF – Chief of Defence Forces

CIA – Central Intelligence Agency

CPA – Comprehensive Peace Agreement

DDR – Disarmament, Demobilisation and Rehabilitation

DPKO – Department of Peacekeeping Operations

DRC – Democratic Republic of Congo

EAC – East African Community

EDF – Equatoria Defence Forces

EPRDF – Ethiopian People's Revolutionary Democratic Front

EPLF – Eritrean People's Liberation Front

FAA – Federal Aviation Administration

FBI – Federal Bureau of Investigation

Gen. – General

GoS- Government of Sudan

GoSS – Government of Southern Sudan

GoU – Government of Sudan

HIV – Human Immunodeficiency Virus

HSM – Holy Spirit Movement

HRW – Human Rights Watch

IC – Intelligence Community

ICC – International Criminal Court

ILC – International Law Commission

INGOs – International Non-Governmental Organisations

Intel – Intelligence

ISI – Inter-Services Intelligence

JIF – Joint Integrated Force

JIF – Juba Initiative Fund

JMAC – Joint Mission Analysis Centre

JOC – Joint Operations Centre

LDUs – Local Defence Units

LOAC – Laws of Armed Conflict

LTTE – Liberation Tigers of Tamil Eelam

LRA – The Lord's Resistance Army

Maj. Gen. – Major General

MI – Military Intelligence

MONUC – UN Mission in Congo

NAS – National Salvation Front

NCP – National Congress Party

NORAD -Norwegian Agency for Development

NRM/A – National Resistance Movement/ Army

OCHA – Office of the Coordination for Humanitarian Affairs

NSA – National Security Agency

Ops – Operations

PHMC – Political Military High Command

RECCE – Reconnaissance

RTF – Regional Task Force

SAF – Sudan Armed Forces

SPLM/A – Sudan People's Liberation Movement/ Army

SSLA – Southern Sudan Legislative Assembly

SSUF – South Sudan United Front

TEW – Terrorism Early Warning Group

TRG – Training

UNLA – Uganda National Liberation Army

UNRF11 – Uganda National Resistance Front 11

UPDF – Ugandan People's Defence Forces

UN – United Nations

UNMIK – The United Nations Interim Administration Mission in Kosovo

WNBF – West Nile Bank Front

# Important Historical Timelines

1947- Juba Conference to decide the fate of Southern Sudan

1953- Sudan given self-Rule by the British Colonial Administration

1955- Torit Mutiny protesting Arab domination

1956- Sudan Independence from Britain

1962-Uganda Independence from Britain

1966- Prime Minister, Milton Obote overthrew Buganda King Mutesa

1969- Nimeiry became President of Sudan through military takes over

1972- Addis Ababa Peace Agreement between the Government and the ANYANYA Fighters

1971- General Amin Dada overthrew Milton Obote

1979, Professors Yusuf Lule and Godfrey Lukongwa Binaisa had short tenures.

1980- Uganda held elections and Obote became the winner in a disputed election

1982- Nimeiry redivide the Southern Region

1983- Nimeiry introduced Sharia Laws

1983- Sudan People's Liberation/Movement/Army was formed

1985- Obote's army general, Tito Okello Lutwa overthrew Obote

1986- NRA/NRM took control of power in Uganda

1986- First revolt in Northern Uganda

1987-NRA defeated UPDA

1987- HSM formed by the prophetess Alice Lakwena

1987- NRA defeated Alice Lakwena HSM

1988- Joseph Kony Joined HSM

1991- SPLM/A split

1991-NRA launched Operations North on LRA

1994- Josep Kony renamed HSM to become the LRA

1994- LRA established relationship with Sudan

1995- Betty Bigombe Initiative for the resolution of the conflict politically

1996- LRA Abducted Aboke St.Mary's School Girls in Northern Uganda

2002- NRA Operations Iron Fist

2003- Human Rights Watch Report of abuses committed during the LRA War

2003- Sudanand SPLM/A began Negotiations in Kenya

2005-Sudan and SPLM/A signed the Comprehensive Peace Agreement in Kenya

2005 –Dr John Garang de Mabior SPLM/A leadership became Sudan First Vice President

2005- Dr John Garang de Mabior died in the Helicopter. Crash

2005- ICC Charged Joseph Kony and other LRA commanders with War Crime

2006- South Sudan mediate between LRA and Uganda Government

2011- South Sudan became independence from Sudan

# Chapter 1

<hr>

# The LRA

## Introduction

If you ask a typical person in Uganda, South Sudan, the Democratic Republic of Congo (DRC), the Central Africa Republic (CAR), and perhaps other places in the Horn and Eastern African regions what they know about the LRA, they will answer that it is a brutal movement led by Joseph Kony. Vinci concluded that the LRA strategically uses fear as a force multiplier to further its organisational survival and as a way to fight a political 'dirty war'.[1] Under Joseph Kony, the LRA has been involved in kidnapping, forced marriages, conscription of underage children, and amputation of ears, hands and other human body parts.

---

1    Vinci, A. (2005). The strategic use of fear by the Lord's Resistance Army. Small Wars & Insurgencies, 16(3), 360-381.

According to Seybolt, in 1999, there were twenty-seven major armed conflicts in twenty-five countries around the world[2]. He mentioned Africa as a region where these wars were taking place.[3] The LRA War was among the conflicts named. The LRA started fighting against the Ugandan government in the late 1980s, killing and mutilating thousands of civilians. Their violent campaign displaced over two million people and left thousands of dead.[4] Before it was so named, the LRA was preceded by other northern rebel groups comprised of the same northern Ugandan people; such as the Holy Spirit Movement (HSM) of Alice Ouma Savarino. This chapter presents the beginning and the causes of the war in northern Uganda. Further, it looks at the progression of the northern Uganda rebellion from Alice Lakwena's Holy Spirit Movement (HSM) to Joseph Kony's Lord's Resistance Army (LRA). Finally, it explains how the colonial divide policy contributed to political instability in Uganda.

**The Northern Uganda Crisis**

The crisis in northern Uganda that culminated into civil war was related to the success of the National Resistance Movement/Army (NRM/A) war that brought president Museveni into power in Uganda in 1986. However, civil war scholars have explained causes of civil war within the parameters of greed, grievance, motivation, and opportunity.[5] War is defined as sustained, coordinated violence between political

---

2        Seybolt, T.B., 2000. Major Armed Conflicts 1999. SIPRI yearbook.

3        Seybolt, T.B., 2000. Major Armed Conflicts 1999. SIPRI yearbook.

4        Godfrey Olukya. Brutal LRA rebel group in crisis. The African Report. Wednesday, 25 February 2015.

5        See Jack S. Levy and William R. Thompson. (2010). Causes of War. Wiley-Blackwell, p.7.

organisations.[6] Whether it was greed, grievance, motivation or opportunity that fuelled the Northern Uganda Civil War, the reality is that Museveni's presidency was received with mixed reactions by the people of Uganda. While the people of western Uganda celebrated because their son Museveni had come to power and, therefore, it was "their time to eat", the people of northern Uganda went to war because they had lost the power they had controlled for decades. The quest for different ethnic groups to get one of their own in power is common in post-independence Africa. Kameri-Mbote writes:

> . . . the failure of post-independence national leaders to craft a cohesive national policy resulting in people relying principally on their tribal/ ethnic alliances to access resources, including land. Successive governments, particularly the Kenyatta and Moi ones, have used allocation of public land to reward supporters, gain favours from its supporters or to ensure political patronage. The quest for different ethnic communities to get one of their own in State House has a lot to do with perceived enhanced access to resources that 'their man' will ensure going by past regimes. Consequently, there are very high stakes in each general election for different ethnic communities and their allies.[7]

Indeed, Uganda was never an isolated case. At independence, Uganda was tribalised at the wider context into either Nilotic or Bantu tribes, with Bantu becoming the ruling elite, supplemented by economic power. The Nilotic tribes subsequently became the soldiers charged with safeguarding the Bantu elites.

---

6      ibid, p. 5.

7      See Patricia Kameri-Mbote. (2009). The Land Question in Kenya: Legal and ethical Dimensions. Strathmore University Press, p. 13.

Tribal competition for resources was evident in the Ugandan political setup at independence. As Uganda stepped towards self-rule, there emerged between ethnic groups a politics of who would capture state power as soon as the British left. Uganda became independent on 9 October 1964 amid political, social, religious and cultural crisis.[8] At independence, the King of Buganda, Mutesa II, became the head of government of independent Uganda, with Milton Obote as prime minister. However, this arrangement did not last for long, as Mutesa would be overthrown and forced into exile. The political dichotomy of Uganda is similar to that of Sudan. Like in the case of Sudan, where colonial administration created disparities in development, the British colonial administration created a landlord's policy in Uganda, with the Buganda region at the centre of this policy and being its main beneficiary. In Sudan, the colonial administration gave all powers of the country to the north. The Buganda ownership of land gave them control over the producers on land and, thus, over the agricultural surplus product.[9] Politically, unlike Kenya and Tanzania, Uganda did not emerge after independence with one ruling political party. Instead, two separate semi-tribal parties emerged: the Buganda and the non-Buganda.[10] There was also division along religious lines, with the organised religions becoming highly politicised.[11]

The colonial administration introduced a policy of tribal ideological

---

8    See Yoweri Kaguta Museveni (2016), Second Edition. Sowing the Mustard Seed: The Struggle for freedom and Democracy in Uganda. Moran EA Publishers. P.256, Mahmood Mamdani (1976). Politics and Class Formation in Uganda. London: Heinemann. p. 120.

9    Mahmood Mamdani (1976). Politics and Class Formation in Uganda. London: Heinemann. P 120.

10    Ibid, p. 229.

11    Ibid, p. 284.

organisation, which Buganda exploited for their benefit.[12] This policy defined or regarded the people of northern Uganda as fit for soldierly work and the people of Buganda as fit for ruling and entrepreneurship. This policy translated itself into the development of the Buganda region in many aspects. Kampala, the capital of Uganda, being in the Buganda region, was in the heart of economic development. With this colonial classification in mind, the Buganda region was a preferred candidate for leadership as Uganda prepared for independence.[13] However, this policy was short-lived, as things changed at independence. Following independence, the northern people, through their military officers, utilised their state's coercive apparatus to get into state power by force, a pattern that continued for decades until President Museveni's ascension to power in 1986 following NRM/A's victory. As Atkinson emphasises:

> *The colonial pattern of [the] northern-dominated army continued for nearly a quarter of a century after Uganda gained its political independence in 1962. This would change following the ascension to power of the current president, Yoweri Museveni, and current National Resistance Movement (NRM) government in January 1986, after a five-year bush war that defeated a predominantly northern army.[14]*

The ascension of Museveni to power manifested itself in widespread violence, especially in northern Uganda. The people of northern

---

12      Ibid.

13      Ronald R. Atkinson (2009). From Uganda to the Congo and Beyond: Pursuing the Lord's Resistance Army. P. 4.

14      Ronald R. Atkinson (2009). From Uganda to the Congo and Beyond: Pursuing the Lord's Resistance Army. P. 4.

Uganda advanced their claim of marginalisation as one of causes of the conflict. Sylvester B Maphosa elaborated this point as follows:

> *In response, a number of northern factions have sought to contest the NRM/A for State power, with the narrative of marginalisation of northern Uganda as the mobilising dynamic. In 1986, Alice Lakwena initiated a movement, part of which later became the LRA, reflecting Acholi opposition to Museveni's NRM/A. This opposition was based on both realistic and non-realistic structural grievances. After Alice, Joseph Kony (cousin to Alice) assumed leadership of remnants of the movement and established LRA as the main local rebel group in northern Uganda, drawing support mainly from sections of the Acholi.*[15]

Thus, the causes of the LRA War in northern Uganda were deeply rooted structural problems. Many structural causes of conflict emerge from the nature of political governments, the relationship between the state and its citizens, the legitimacy of the government, and the state's ability to provide basic services. The LRA War, therefore, can be analysed within the Kenneth Waltz and Singer framework, which includes both factors associated with the government and factors associated with society.[16] Factors associated with government include variables like the institutional structure of the political system and the nature

---

15     Sylvester B. Maphosa. The Lord's Resistance Army: A Review of African Union Regional Efforts to Eliminate the Resistance in Central Africa, in A Comprehensive Review of African Conflicts and Regional Interventions 2016. P. 212-264.

16     This framework goes back to Kenneth Waltz's book Man, the State, and War. (1959), which identified three images of war, and Singer (1961), three levels of analysis.

of the policy-making process.[17] Factors associated with society include variables like the structure of the economic system, the influence of economic and noneconomic interest groups, the role of public opinion, and political culture and ideology[18].

Although disputed by scholars such as Wright,[19] the societal hypothesis suggests that some cultures are more warlike than others.[20] It is not clear whether the colonial administration in Uganda was influenced by this belief. At the advent of independence, Ugandan tribes were categorised into two groups: those who could rule and those who could do soldierly work. Acholi and many other northern Ugandan tribes were given the role of soldierly work. With hope to restore the power they had lost to Museveni; the people of northern Uganda rushed to their stronghold to form resistance movements. Thus, the Lord's Resistance Army (LRA) was born, and insecurity increased. However, as Chandra et al argues, there is a confluence of several causes for a given conflict, even if one cause can be identified as more prominent.

Joseph Kony, whose whereabouts remains unclear today, inherited Alice Lakwena's Holy Spirit Movement (HSM) following her defeat by the NRA in 1987. The HSM was one of many rebel movements in Uganda formed at the time of Museveni's takeover of the seat of

---

17      Jack S. Levy and William R. Thompson. (2010). Causes of War. Wiley Blackwell. P.14.

18      Ibid.

19      See Wright, Quincy (1965). A Study of War. Chicago. Chicago University Press.

20      Jack & Thompson, p. 15.

presidency in Uganda in 1986.[21] As Hemmer states:

> *Northern Uganda has been the scene of armed conflict ever since President Yoweri Museveni, a southerner, took power in 1986 by overthrowing a military regime dominated by the Acholi, the largest ethnic group in Uganda's northern districts of Gulu, Kitgum and Pader. In the wake of this coup, several protest movements emerged in the north, challenging the newly established leadership due to fears of political and economic marginalisation. The LRA was born out of these movements and has used guerrilla warfare against the national army, the Ugandan People's Defence Forces (UPDF), and terrorised local communities in the north since 1987.[22]*

The magnitude of rebellion against NRM/A has also been acknowledged by Museveni himself, observing that:

> *Uganda has become an endemically insecure area . . . in our time, the insecurity was caused by terrorists supported by the Sudan and by cattle rustlers . . . The terrorist's groups were many. The significant ones were Kony's group (the LRA) and the ADF.[23]*

---

21      See Kasaijja Philip Apuuli. 'The International Criminal Court and the Lord's Resistance Army Insurgency in Northern Uganda, in Alfred Nhema and Paul Tiyambe Zeleza. Eds. (2008). The Resolution of African Conflicts: The Management of Conflict Resolution and Post Conflict Reconstruction. James Currey. P. 52.

22      Hemmer, J. (2008). The Lord's Resistance Army: tackling a regional spoiler. Clingendael Institute, p. 1.

23      Yoweri Kaguta Museveni (2016), Second Edition. Sowing the Mustard Seed: The Struggle for freedom and Democracy in Uganda. Moran EA Publishers. P.256.

The objective of HSM was to fight and dismantle the National Resistance Movement/Army (NRM/A) of President Yoweri Kaguta Museveni, who they believed to have forcefully taken over the leadership of Uganda on 26 February 1986 after their six years of protracted rebellion in Uganda. Museveni rebelled and formed the NRM/A to fight and liberate Uganda from what he called disoriented and reactionary politics that had engulfed Uganda before, during and after colonialism.[24] Tim agrees and suggests that the root causes of the northern Uganda conflict are traceable to the colonial period, when the British administration classified jobs on ethnic and regional lines.[25] Following Uganda's attainment of independence in 1963, it is perceived that the British elevated Baganda to power, whereas the Acholi and other northern people of Uganda were left at the periphery, thus creating general mistrust and grievances between tribal groups in Uganda.

**Alice Lakwena Holy Spirit Movement: The Foundation of the LRA**
Lakwena regrouped the remnants of the defeated Uganda National Liberation Army (UNLA) led by Erica Odwar[26] under a new name, the Holy Spirit Movement (HSM). The HSM launched a series of attacks on UPDF, but unfortunately, they all failed. The hasty defeat of UNLA and HSM can be attributed to the little knowledge that the leadership of both organisations had. Lakwena, for example, instead of focusing on military strategy, emphasised spiritual aspects that

---

24      Yoweri Kaguta Museveni (2016), Second Edition. Sowing the Mustard Seed: The Struggle for freedom and Democracy in Uganda. Moran EA Publishers. P.146.

25      Ibid.

26      Erica Odwar was killed in a battle field with NRA at Kilak in 1987. For more details see Museveni. P. 258.

contradicted military strategy. On this, Museveni says, "Alice Lakwena, with no military knowledge and being a slave to mysticism, decided that she had to reach Kampala."[27]

As is illustrated above, it is undoubtable that the objective which led to the formation of HSM centred on the restoration of power that the people of northern Uganda had lost to Museveni. Chandra et. al argued:

> *When the interests or agendas of two or more groups are at odds, such as over access to resources, political opinion, or religious beliefs, or when two or more groups perceive themselves as opponents because of ethnic or racial origins, conflict becomes inevitable.*[28]

This point is well presented by Apuuli who wrote: ". . . all the insurgencies in northern Uganda, including that of the LRA, can be explained as an attempt by the people of that region to regain the power that they lost in January 1986 following the victory of Museveni's NRM."[29] Lakwena was defeated but left a legacy behind. Her legacy brought about the formation of the LRA. Lakwena mobilised a remarkable force during her leadership of the HSM. As Matthew Green notes, "Alice Lakwena was able to mobilise a rebel army of ten thousand fighters in a short possible time before the NRA defeated her in

---

27      Yoweri Kaguta Museveni (2016), Second Edition. Sowing the Mustard Seed: The Struggle for freedom and Democracy in Uganda. Moran EA Publishers. P.146.

28      Sriram, Chandra L., Olga Martin-Ortega and Johnna Herman (2009). War, Conflict and Human Rights: Theory and Practice, Routledge.

29      Kasaija Philip Apuuli 'The International Criminal Court and thr Lord's Resistance Army Insurgency in Northern Uganda' in Alfred Nhema & Paul Tiyambe Zeleza (2008). The Resolution of African Conflicts: The Management of Conflict Resolution & Post-Conflict Reconstruction. James Currey. P.52 .

October 1987."[30] This force became the foundation of the LRA.

Lakwena's defeat forced her, eventually, to surrender leadership of the movement to her father, whose leadership was short-lived. Following her retirement from the rebellion, Lakwena sought refuge in Kenya, where she died. But why did Lakwena quickly abandon the movement she founded and built when many of the fighters of the HSM and its successor, the LRA, saw themselves to be fighting for the salvation of their people, the Acholi, whom they believe to have been marginalised, abused and excluded from Uganda's development by what they termed the oppressive regime of President Museveni?[31] Amilcar Cabral notes, "[t]he people fight and accept sacrifices demanded by the struggle in order to gain material advantages, to live better and in peace, to benefit from progress and for the better future of their children."[32] Was it really because of the defeat? Was her life in danger from her comrades following her defeat? Was she finding it difficult to run affairs in a predominantly male organisation? These questions and many more are for other people to study. Following the defeat that forced her to take refuge in Kenya, an attempt was made to let Lakwena's ageing father succeed her, but this failed due to his advanced age. This led him to surrender the movement's leadership to Joseph Kony.

---

30      Green, M., 2009. The wizard of the Nile: the hunt for Africa's most wanted. Gardner's Books. P.4.

31      Simonse, S., Verkoren, W., & Junne, G. (2010). NGO Involvement in the Juba peace talks: the role and dilemmas of IKV PAx Christi. The Lord's Resistance Army: Myth and Reality, 223-242.

32      See Paul Nugent (2004). Africa Since Independence. Palgrave Macmillan. P. 260.

## The Birth of the LRA

As part of his reform agenda, Kony renamed HSM the Lord's Resistance Army (LRA). Soon after his takeover, Joseph Kony and the LRA posed a serious challenge, not only to the government of President Museveni but also to international security, as the LRA's operations went beyond Uganda's border into South Sudan, the Central Africa Republic and the Democratic Republic of Congo, prompting them to share intelligence. By November 2011, it was reported that the LRA remained a deadly threat to civilians in these countries.[33] Like the HSM, which first established its stronghold in Acholi land in northern Uganda, the LRA had its bases in the Acholi districts of Gulu, and Kitgum. This was due to obvious reasons of sympathy, since both HSM and LRA were champions of Acholi people's rights, including restoration of the leadership seat that they had lost to Museveni in 1986.

The LRA presented clear objectives centred around the welfare of the people of northern Uganda, but what remains to be established is why the LRA, under Joseph Kony, became notorious and famous for its atrocities in northern Uganda against the people Kony claimed to be fighting for. The LRA's atrocities in northern Uganda included abduction of children for use as sex slaves and porters, to chopping off lips, ears, noses and other body parts.[34] Some researchers have said that Joseph Kony, a self-proclaimed prophet who claimed to take orders from the Holy Spirit, wanted to established theocratic rule in Uganda

---

33      International Crisis Group. The Lord's Resistance Army: End Game? Africa Report No. 182-17 November 2011.p.i.

34      Such atrocities are well documented, see for example Green, M., 2009. The wizard of the Nile: the hunt for Africa's most wanted. Gardners Books. P.9.

to be based on the ten Christian Biblical commands.[35] However, the LRA manifesto stated that while many members of the LRA were practicing Christians, they did not intend to become fundamental Christians.[36]

Joseph Kony, formerly a Roman Catholic catechist, leads the LRA to this date. Unconfirmed information suggests that he is hiding and roaming between Sudan's western region of Darfur, the Central African Republic and the Democratic Republic of Congo. Following mounting pressure on his forces resulting from intelligence-sharing between the US, Uganda, Sudan and South Sudan, Joseph Kony and his forces relocated to South Sudan's western Equatoria border to escape approaching UPDF and SPLA operations, following the failure of the Juba peace talks.

The failure of Juba peace talks strengthened the need for the SPLA to join the UPDF war against the LRA, as the commitment to neutrality in the role of mediator, which previously hindered SPLA involvement, was now removed. With South Sudan mediating the Uganda warring parties peace talks and its forces SPLA fighting alongside UPDF, it would have been accused of impartiality which would in turn justify the LRA's objection mediation by South Sudan. The LRA has always been, from the very beginning defined by South Sudan as an enemy because of its link with the Sudan government which the SPLM/A fought for many years before its independence in 2011. The LRA continued to disturb South Sudanese peace in their villages and this made South Sudan even more adamant to fight the LRA. So the

35    See for example Green, M., 2009. The wizard of the Nile: the hunt for Africa's most wanted. Gardners Books. P.9.

36    Finnstrom, Sverker, 2003. Living with Bad Surroundings: War and Existential Uncertainty in Acholiland, Northern Uganda. Uppsala University Press.

LRA refusal to sign peace deal with Uganda was now an invitation for the South Sudan government to join war on LRA. With SPLA's full participation in the war, the UPDF had more advantage to win war against the LRA. Indeed, within a few months, the LRA war in eastern Uganda and South Sudan was over. The LRA was forced into the DRC where are assumed to be hiding to this date.

There has been confusion over LRA objectives, as some writers suggest that Joseph Kony has been fighting for the establishment of theocratic rule in Uganda.[37] These writers have been dismissed, as the brutal conduct of the LRA does not conform to religious principles. Tim Allen, in his book *Trial Justice*,[38] highlighted the LRA's objectives to obtain state power and stop the abuse of power by the NRM/A in Acholi land.

Joseph Kony's greatest mistake was killing and mutilating innocent civilians who refused to take part in war. Museveni says, "His greatest crimes were to kill and mutilate non-combatants and abduct unwilling fighters."[39] By going with Allen's highlighted LRA objectives, it is fair to say that the LRA has fallen short of its vision to stop government abuse of the people of northern Uganda. The LRA's character betrayed the cause of its own war, and it is therefore not a just war. It is contradictory to fight for people's rights on one hand and abuse them on the other. Throughout its history, the LRA has continuously attacked the people of northern Uganda and Southern Sudan, practicing on them all sorts of inhumane treatment, including cutting of their genitalia.

---

37      De Temmerman, E., 2001. Aboke Girls, Children Abducted in Northern Uganda. P.15.

38      Allen, Tim (2008), Trial Justice: The Criminal Justice and the LRA, p.53.

39      Museveni, p. 267.

Simonse and Junne say, "[t]he LRA is widely known for its abduction of people, the mutilation of victims and its very brutal attacks upon civilians."[40]

The LRA can be contrasted with HSM when coming to their nature of operations. While the LRA is known for its brutality, Alice Lakwena's Holy Spirit Movement (HSM) the predecessor of the LRA understood that waging war was based on the moral values that the HSM was founded on.[41] Lakwena founded her movement in August 1986, just few months after the NRA installed itself as a government in Kampala. The HSM, like the LRA, was founded to fight for the rights of the Acholi people first, then to liberate Ugandans from Museveni's tyranny. It is puzzling to see that freedom fighters are inhumane and abusive towards the people they claimed they were fighting for. These abuses manifested themselves in the way the LRA treated civilians and captives. Comparatively, the HSM cautiously or incautiously adhered to the Laws of Armed Conflict (LOAC).[42] This is understandable because Lakwena established HSM as an attempt to reconstitute the moral order based on the formulation of an alternative theory of social tensions and power relationships, with the idiom of religion and ritual or, as Behrend calls it, 'edification by puzzlement'. Waging war was thus understood as an ordeal but a nonetheless necessary instrument in the process of cleansing or purifying, rather than separating from the unjust, to create a healed rather than suffering community.[43]

---

40      Ibid.

41      Heike Behrend (1999), Alice Lakwena and the Holy Spirits: War in Northern Uganda 1986-1997.

42      See Museveni, p.267.

43      Frank Van Acker. (2004). Uganda and the Lord's Resistance Army: The New Order no one ordered African Affairs, Vol. 103, No. 412 (Jul., 2004), pp. 335-357.

## The LRA's Political Objective

Many theories have emerged with regards to the LRA's political objectives, some describing Joseph Kony as a madman with no political objectives or goals.[44] However, Hobbes, decades ago, suggested that men compete with each other for the same things. It is this competition that inevitably results in conflict.

Debating the LRA's political objectives is not the aim of this book. What is known, though, is that the LRA war began in northern Uganda, and for this very reason, it became well known as the Northern Uganda Civil War. The Northern Uganda Civil War, or the LRA War, began in 1986 to pursue several objectives; including the restoration of the northern Uganda leadership, which had been overthrown by Museveni and his rebel organisation, the National Resistance Movement/Army (NRM/A). The LRA first began when a great number of Acholi soldiers who had lost their jobs in the Ugandan Army, retreated north following the overthrow of the Tito and Basilio Okello government by Museveni on 25 January 1986. Both Okellos were, by birth, from the northern districts of Uganda. They briefly ascended to the state house in Uganda in their bloodless coup against Obote in July 1985.

The Northern Uganda War, led by Joseph Kony, had different dimensions, some of which are political, social, ethnic, and cultural. The war also brings in an element of ethnic nationalism, which is focused on the unity of the people of northern Uganda; primarily the Acholi and the Langi. Nationalism in general is a phenomenon whereby people of similar persuasion or origins, and identities derived thereof, express and act together in pursuit of maintenance of their common

---

44      See Dunn, K.C.2004. Uganda: The Lord's Resistance Army. Review of African Political Economy, 31(99), pp 139-142.

aims and interests, resulting in political manifestations, whether psychologically or territorially defined.[45] Francis Fukuyama echoes this, arguing that '. . . because they demanded recognition of the dignity of the group in question, they turned into political movements that we label nationalism. . .'[46]

Just like the NRM/A war, the LRA war was rooted in the Uganda post-independence dysfunctional state,[47] which was dominated by the northern leadership since independence from British colonial administration. Milton Obote himself, from the northern region of Uganda, took over the position of the office of prime minister at independence. He held the position until he was overthrown by another northerner from Western Nile, General Iddi Amin. Amin remained in power until he was overthrown by Obote with the help of President Nyerere of Tanzania in 1982.

The power shift in Uganda from the northern region was only signalled when Museveni's National Resistance Movement (NRM) took over the country's leadership in February 1986. Emma Leonard writes:

> . . . the coming to power of the NRM also signalled a shift in power
> within the country. Museveni and most of the original members of the
> NRM were from the southwest of Uganda and were ethnically Ankole.
> This was the first time since independence that the president of Uganda

---

45      Amoah, M. (2011). Nationalism, Globalization, and Africa. USA: Palgrave Macmillan. p. 1.

46      Fukuyama, Francis (2018). Identity: Contemporary Identity and the Struggle for Recognition. London: Profile Books, p. 59.

47      For better understanding of the meaning of dysfunctional state, see Ghani, A. and Clare Lockhart (2008). Fixing Failed States. Oxford, Oxford University Press, p. 65.

*had not been from one of the northern regions and the first time that the Ugandan army had not been made up of substantial numbers of northerners. As part of their colonial rule in Uganda, the British had instituted a policy of divide-and-rule. Thereby, Bugandans, based in the centre and south of Uganda, staffed most of the bureaucracy, while the army was predominantly manned by Acholi and Langi from the north.[48]*

Woodward also notes that:

*Museveni came from western Uganda, but in broad terms, his rise to power appeared to be that of the southern part of the country, which had for so many years been under the sway of men from the north.[49]*

Prudence Acirokop echoed this and argued that "Museveni's power base was a largely southern army replacing the northern political and military rule known since independence."[50] Unfortunately, the northern Uganda war became one of the worst civil wars recorded in terms of human rights violations perpetuated by the LRA. The gross human rights violations made the LRA an unpopular organisation both locally and internationally.

The Oxford Dictionary of Law defines war as "the legal state of affairs

48    Emma Leonard. The Lord's Resistance Army an African Terrorist Group? Perspectives on Terrorism, Vol. 4, No. 6 (December 2010), pp. 20-30 Published by: Terrorism Research Initiative.

49    Woodward, P. Uganda and southern Sudan 1986-9: new regimes and peripheral politics. In Hansen, H. B and Michael Twaddle (1991). Edits. Changing Uganda. London: James Currey. 178.

50    Prudence Acirokop. Accountability for Mass Atrocities the LRA conflict in Uganda. LLD Dissertation, University of Pretoria August 2012.

that exists when states use force to vindicate rights or settle disputes between themselves."[51] According to the Dictionary of International Relations, a civil war is a protracted internal violence aimed at securing control of the political and legal apparatus of a state.[52] This definition introduced the concept of the state, and therefore, it is importance to define it as well. In accordance with Article 1 of the Montevideo Convention, "the state as a person of international law should possess the following qualifications: a) a permanent population; b) a defined territory; c) a government; and d) capacity to enter into international relations with other states."[53]

There have been many civil wars around the world. These include the English Civil War (1642–1651), the American Civil War (1861–1865), the Spanish Civil War (1936–1939), the Sudan Civil Wars (1955–1972; 1983–2005), the Ethiopian Civil War (1974–1991), the Somali Civil War (1989-present), the Congolese Civil War (1665-1709), and the Algerian Civil War (1954-1962) to name few. However, what distinguishes these wars from the northern Uganda war is the level of gross human rights abuses. The LRA abuses went beyond Ugandan borders, crossing into South Sudan, the Democratic Republic of Congo, and the Central African Republic. As Kristof Titeca and Theophile Costeur state:

*During the last decade, the Lord's Resistance Army (LRA) became a regional problem in the border area of the Democratic Republic of*

---

51      Martin, E.A. Ed. (2006). A Dictionary of Law. Sixth Edition. Oxford: Oxford University Press. p. 570.

52      Evans, G. and Newnham J. (1998). Dictionary of International Relations. England: Penguin Books. P.64.

53      Article (1) Montevideo Convention 1933.

*Congo, South Sudan, and the Central African Republic, involving multiple national and international actors.*[54]

## The LRA War Strategy

The LRA operated against the guerrilla principles of a fish in a river. The LRA despised the famous quote of Chairman Mao, thus putting it into trouble. Mao suggested that 'guerrilla force is like the fish in the river'. This principle has influenced many guerrilla forces globally with exception of the LRA so to speak. Indeed, the way the LRA treats the people it aspires to liberate, suggests that it was never a fish in a river. This was and still is a serious bottleneck issue for the international community, which believes in human dignity. Therefore, the big question that remains to be answered by the LRA and its leaders, is why it behaved the way it did towards the civil population, when rebel movements are expected to take Chairman Mao of China's advice, who described guerrilla force relations with civil population like the fish in the water. The LRA's mass abduction of children, young men and women seemed to be a general strategy to fight war in northern Uganda. Available records that suggest by 2007, the LRA had displaced 1.5 million people in Uganda[55]. Further, in its 15 years of existence, reports suggest that a total of 30,839, mostly children, were registered as having been abducted by the LRA from 1986-2001[56]. This made LRA notorious for terrible acts instead of any positive ones. As Branch

---

54    See Kristof Titeca and Theophile Costeur. (2014) An LRA for Everyone: How Different Actors Frame the Lord's Resistance Army. African Affairs, Oxford University Press, 114/454, 92–114.

55    See John Prendergast. What To Do About Joseph Kony. Enough Strategy Paper No. 8. October 2007, p. 3.

56    Approximately 2,067 people were abducted every year.

states, "the LRA has become famous for massacres, maiming, and the forced recruitment of thousands of Acholi, many of them children"[57].

Such acts do not make civilized people happy at all. They are intolerable. In consideration of this and with a hope of bringing abduction to an end, the ICC issued in July 2002 the arrest warrant on Joseph Kony, the leader of the LRA, and charged him with thirty-three counts of war crimes and crimes against humanity. These counts included LRA violations across Uganda borders. Evidence of these violations are found in UN reports. In 2009 alone, the United Nations reported that the LRA had killed more than 1,500 people, abducted more than 1,800 and displaced hundreds of thousands of people in the Central African Republic, Democratic Republic of Congo and Southern Sudan[58]". In a separate report UNICEF, reported that 5,923 were abducted as children, while 7,327 were adults. UNICEF further believed that about 5,555 children remained in Kony's captivity at that time. In 2008 a report from Southern Sudan revealed Kony forces to be composed 70% of children abductees"[59]. Even after all this, there is uncertainty with LRA atrocities. Resolve Uganda, a US based advocacy group, estimates that the LRA still continues to have the capacity to commit brutal atrocities against civilians unless otherwise[60]. These range of atrocities suggest nothing but that the LRA is an organization of oppression of the young and old.

Research conducted in Northern Uganda into the organization and tactics of the LRA revealed that it is an organization that uses violence

---

57      Adam Branch, p. 180.

58      AFP February 10, 2009.

59      Journal of Human Rights Quarterly, Volume 30, Number 2, May 2008.

60      Resolve Uganda report by the UN Radio Okapi, January 13 2009.

to reward its members [61]. Further, it revealed that in the absence of economic and social endowments with which to reward membership, the LRA abducts potential fighters from the civilian population in a way that is analogous to resource exploitation. The report suggests that the LRA has not only used force to recruit combatants into its army but also to accrue rents in the form of military aid from the Sudanese government. The research concludes that the group's leader, Joseph Kony, acts as rentier, by using this self-sustaining conflict to maintain his only source of political capita – the continued threat posed by the LRA. As a result, Resolve Uganda was eager for a greater regional and international involvement to settle the dispute once and for all. According to them the UPDF, which has been fighting LRA, has shown for over 20 years its inability to defeat the LRA and with some analysts suggesting that the UPDF has used its war against LRA to equip itself and that the UPDF has become a significant economic player through its plundering activities and corruption. Despite these criticisms and scepticisms on the way in which the UPDF has handled affairs, it is fair enough to give UPDF some credit.

The LRA has never disputed its connection with Sudan but explained that its connection with them came after they were invited by the Government of Sudan to be friendly troops[62]. According to Salehyan countries in the Horn of Africa tend to seek solutions to their problems by providing sanctuary to rebel groups from neighbouring states, and they intervene in the affairs of their neighbours. to

---

61      Bevan, J (2007) The Myth of Madness: Cold Rationality and 'Resource' Plunder by the Lord's Resistance Army, Civil Wars, 9:4, 343-358.

62      Mareike.p. 18.

counteract perceived threats to domestic interest[63]. Dor explains that:

> *In April and May of 1994, an exceptionally intense struggle raged between the SPLA and the SAF, which was supported by local militias from the Lotuko, Mundari and Toposa regions, as well as by the Lord's Resistance Army (LRA) rebels from Uganda.*[64]

The LRA had a reason to seek friendship with Sudan as the SPLA had sought relationship with the Government of Uganda for reasons of survival. Making alliance was part of the military strategy the SPLA and the LRA adopted. This was never strange since as it is already argued that the enemy of the enemy is a friend. This was a game the SPLA and the LRA played. Unfortunately, this game turned unintended enemies to actual enemies, as the case of the SPLA and the LRA demonstrates. Indeed, the LRA became a threat to the SPLA deployments in Eastern Region on South Sudan Uganda border. From the1990s to the time the SPLM/A became the government of South Sudan, the CPA then ended 22 years of hostilities between the northern and southern Sudan. The SPLA, on the other hand collaborated with UPDF in fighting the LRA. In 1994 Uganda supplied SPLA with warfare and machines, which the SPLA used to push away SAF and LRA from South Sudan-Uganda border. Museveni explains:

> *In 1994, we had given good equipment to the SPLA.... The SPLA and these volunteers, using our equipment, inflicted a defeat on the Sudan Army and pushed them from near the Uganda border on the*

---

63      See Wassara, p. 104.

64      Dor, p. 228.

*side of Ngomoromo, out of Parjok, out of Palutaka and Magwi all the way to Aruu Junction.[65]*

These bases had also hosted LRA forces. Thus, it was also a defeat to LRA and it was from here, that the LRA lost the battle ground against Museveni and the NRM government in Kampala. Such kind of alliance between Sudan government and the LRA on one side, against SPLA and Uganda government on the other, is described by Carl Von Clausewitz to be a strategy of the use of the engagement for the purpose of war in order to further a political end goal[66]. Museveni alliance with SPLA succeeded to force out the LRA of South Sudan Uganda border, deep into South Sudan. By 2006 the LRA had relocated into DRC, CAR and Sudan western Darfur region, where it continued to disturb people's normal lives there by looting, raping, killing and the illegal poaching of wild animal in the parks.

## Conclusion

The LRA is one of a series of rebellions that emerged in northern Uganda to challenge the government of Yoweri Museveni, who successfully fought a short guerrilla war in Uganda from 1980-1986. Unlike many rebels' movements who survive through public relations among the civil population, the LRA used force to recruit and bring people into his rank and file. In 1996 the LRA abducted secondary school girls of St. Mary's Aboke. This is one of the many unwanted behaviours of the LRA that Uganda dedicates for remembrance every October 10. Because of this the LRA lost legitimacy in Uganda and thus, becoming enemy of the region and the world.

---

65      Museveni, p. 268.

66      See Aaron Edwards. Strategy in War and Peace. P. 16.

# Chapter 2

⚬⚬⚬⚬⚬⚬⚬⚬⚬⚬⚬⚬⚬⚬⚬⚬⚬

# Human Rights in the LRA War

## Introduction

Sociologist Emile Durkheim long ago noted that the individual is dominated by a moral reality greater than himself.[67] The moral reality that Durkheim expressed probably never explained the conduct of Joseph Kony, for unfortunately it contradicted natural law theory. The theory of natural law generally advocates that some laws are basic and fundamental to human nature and are discoverable by human reason without reference to specific legislative enactments or judicial decisions. But is Joseph Kony not able to reason, then? Did he not deduce that killing those who were not taking part in war, was a bad

---

67      Durkheim, E. and Suicide, A., 1952. A study in sociology. London: Routledge & K. Paul.

thing? Christian philosophers such as Saint Thomas Aquinas perpetuated this idea, asserting that natural law was common to all peoples—Christian and non-Christian alike—while adding that revealed law gave Christians an additional guidance for their actions.

The defeat of the HSM by the National Resistance Army (NRA) compelled those remaining from the HSM to regroup and reorganise themselves into a new movement, the Lord's Resistance Army (LRA). Joseph Kony became the leader. However, how Kony ascended to leadership is not clear. What is known, though, is that over the course of Kony's leadership, the LRA abducted more than 20,000 children to use as soldiers, servants, or sex slaves, according to UNICEF, leading to violence that displaced more than 2.5 million people.[68] It is yet to be established what motivated Joseph Kony to concentrate on killings, raping, abductions etc . It is believed that the LRA largely ruled through fear. At the peak of its power, in the second half of the 1990s and the early 2000s, the LRA had about 3000 combatants to fight nearly 100,000 government soldiers, so maximizing its threatening appearance and unpredictability was crucial to its survival. The movement used unpredictable and extreme violence to control the population. Photography was part of this strategy.

Adhering to its policy of no mercy for its opponents, the LRA found itself perpetrating serious human rights violations. This was because the LRA resorted to contextualising violence and the use of terror as a means of mobilisation and population control. [69] The LRA became one of the worst human right abusers on earth. However, Uganda

---

68      Cooper, H. A Mission to Capture or Kill Joseph Kony Ends, Without Capturing or Killing, The New York Tines.15 May 2017.

69      Ibid.

military has been implicated in the violations too. These violations are further discussed in details as below.

## The LRA Human Rights Abuses

The LRA human rights abuses transnational international borders and have crossed into South Sudan, DRC and CAR, thus, prompting international outcry and condemnation from the countries affected and their sympathizers. The LRA abuses are in many forms. As of the LRA recruitment policy, Joseph Kony, the leader of the Lord's Resistance Army (LRA), engaged in child abduction to conscript them into his forces. It is observed that the LRA conscripts' young boys and used them as porters and soldiers, and the girls are forcefully married[70]. Kony and his troops perpetrate brutal abuses on the people of northern Uganda.

It is documented that the LRA use systematic killing, abduction, rapping and forced recruitment of civilians as a mean to advance their objective. The LRA's human rights abuses have been reported and documented widely. The latest report was by the French news magazine *Jeune Afrigue*:

> ...*Kony's movement, the Lord's Resistance Army (LRA), began in the mid-1980s with the goal of protecting Northern Ugandans from the newly installed regime of President Yoweri Museveni. But the insurgency turned against civilians, becoming notorious for mutilations and large-scale abductions. By 2006, the LRA had abducted up 38,000 children and 37,000 adults' researchers estimated. Those abducted*

---

70      Quinn, J. R. (2009). Getting to peace? Negotiating with the LRA in Northern Uganda. Human Rights Review, 10(1), 55-71.

*were forced to become fighters or fighters' 'wives'—a euphemism for sex slaves—and household servants.*[71]

This report and many others, were significant in shaping the thinking of the international community towards the LRA. The immediate result of this has been the labelling of the LRA as a terrorist organisation. Thus, as far as the international community is concerned, the LRA's human rights record is apalling, and for this reason, the LRA's leader, Joseph Kony himself, and other top commanders are wanted by the International Criminal Court (ICC) to answer questions regarding these abuses. On 13 October 2005, the ICC issued arrest warrants against Joseph Kony and four of his top commanders, charging them with war crimes and crimes against humanity.

The burden of the LRA war in the northern Uganda region in terms of human rights abuses has been unquestionably bad. Jort Hemmer states:

> *Northern Uganda has unquestionably suffered the greatest burden of LRA activity…Since 1987, tens of thousands of people have been killed, and 1.8 million were displaced at the height of the conflict.*[72]

Enough Project has also echoed the LRA abuses. It writes:

> *The fate of a war that has crossed three international borders, abuses national borders, displaced nearly two million people, and created the*

71    Kristof Titeca How Joseph Kony's notorious Lord's Resistance Army uses photographs as weapons. The Washington Post. 9 December 2019.

72    Hemmer, J. (2008). The Lord's Resistance Army: tackling a regional spoiler. Clingendael Institute, p. 1.

*highest child abduction rate in the world hinges on the fate of one man: Joseph Kony, the notorious leader of the rebel Lord's Resistance Army (LRA).[73]*

According to Charles Baguma, estimates by human rights groups calculate that at the height of the violence, more than two million people were forcibly relocated to internally displaced persons (IDP) camps, and tens of thousands of civilians, including men, women and children, were abducted, while thousands more were killed by the LRA.[74] Baguma argues that many villagers suffered from persistent LRA attacks and that the people of Uganda still remember the 20 April 1995 massacre, when the LRA, after an intense offensive, defeated the Ugandan army and entered the trading centre of Atiak.[75] It is alleged that the LRA rounded up hundreds of men, women, students and young children and marched them a short distance into the bush. After being separated according to sex and age, they were accused of collaboration with the government, and the LRA commander in charge ordered his soldiers to open fire three times on a group of about 300 civilian men and boys, as women and young children witnessed the horror and were told to applaud the LRA's work.[76] Youths were forced to join the LRA to serve as the next generation of combatants and sexual slaves.

---

73      See John Prendergast. What To Do About Joseph Kony. Enough Strategy Paper No. 8. October 2007, p. 1.

74      Baguma, Charles (2012) "When the Traditional Justice System is the Best Suited Approach to Conflict Management: The Acholi Mato Oput, Joseph Kony, and the Lord's Resistance Army (LRA) In Uganda," Journal of Global Initiatives: Policy, Pedagogy, Perspective: Vol. 7: No. 1, Article 3. Available at: http://digitalcommons.kennesaw.edu/jgi/vol7/iss1/3

75      Ibid.

76      Ibid.

Clearly, there is a reason why the LRA has become an international enemy. It is the moral duty of the international community not to stand by and watch while a group of thugs kill people indiscriminately. The LRA's behaviour may be explained in many ways. It is attributable to the changing nature of human warfare that scholars such as Levy and Thompson recognised some years ago.[77] According to Levy and Thompson, war varies in terms of who fights, where they fight, how often they fight and with what intensity.[78] This point is observed by Chandra, Olga and Johanna, who argue that human rights abuses can emerge as a result of violent conflict.[79] Indeed, since it started waging war against the government of President Museveni in the late 1980s, the LRA's attacks were concentrated and directed on the civil population wherever they operated. Frank Van Acker summarises these abuses in the following words:

> *The mutilation and summary execution of non-combatants, the abduction of children and adults for use as foot soldiers, sex slaves and porters have measured the cadence of this conflict with [the] regularity of a metronome.*[80]

These statements give us full insight and magnitude of the LRA's war abuses. Such abuses were committed throughout the conflict

---

77    See Jack S. Levy and William R. Thompson. (2010). Causes of war. Wiley-Blackwell. P.11.

78    Ibid, p. 20.

79    Sriram, Chandra L., Olga Martin-Ortega and Johnna Herman (2009). War, Conflict and Human Rights: Theory and Practice, Routledge. p. 6.

80    See Frank Van Acker (2004). Uganda and the Lord's Resistance Army: The New Order No One Ordered. African Affairs. Vol. 103. No. 412, July, pp.335-357.

in northern Uganda and in South Sudan, especially in Eastern and Western Equatoria states, where the LRA relocated to avoid mounting pressure from Uganda and South Sudan with the help of the US intelligence contingent. The LRA's abuses were widely documented by both local and international non-governmental organisations (INGOs) that were providing humanitarian aid in Northern Uganda and Southern Sudan. The INGOs' documentation of the LRA's attacks on the people of northern Uganda and their brothers and sisters in South Sudan mobilised the international community's opinions against the LRA and created sympathy towards the government of Uganda. This sympathy eventually led the Uganda People's Defence Forces (UPDF), backed by US special forces and African Union (AU) troops, to eventually enter South Sudan to search for Joseph Kony and the LRA.

## The Abduction of St. Mary's College Girls

The case of Aboke is tragic to tell but became a well-known incident due to the international attention given to it. In brief St. Mary's College, Aboke, located in the Lango Apac District of northern Uganda, was operated by Sister Rachele of Combonian missionaries. On 9 October 1996, one hundred and thirty-nine schoolgirls were abducted from their school boarding facilities by the Lord's Resistance Army (LRA). Joseph Kony's war by the time of school raid had just reached a decade since its abrupt rise in 1986. The abduction was condemned worldwide and all the voices called for immediate release of the school girls. The Headmistress Sister Rachele sacrificed her life by commuting between Kony's camp located in South Sudan and meeting highest political and religious authorities globally to make sure that these girls were safe and to be released. Among those she visited to lobby for the release of the abductees were former Pope John Paul II, Kofi Annan former Secretary General of the United Nations, and the presidents of Uganda, Sudan and South Africa.[81] Her efforts were fruitful. One hundred and nine girls were released and were returned to their families as mothers with LRA babies. The other thirty were missing and could not be accounted for by the LRA. It is likely that they were kept as hostage women by the LRA mostly likely by their top commanders.

The story of St. Mary's College school girls is miserable to tell on one hand because it involves human suffering but on the other, it was a turning point in Joseph Kony LRA northern Uganda war for it both mobilised international community and any sympathisers within, against Joseph Kony and the LRA movement. More importantly, the decision taken by the international community to disregard Kony became a lesson for others who were involved in civil wars in Africa and other regions of the world, that abductions and other associated violations were not condonable.

## The LRA War Crimes

Attempts to address serious crimes followed the end of the First World War, when the Versailles Treaty provided for criminal responsibility of German state officials and which were handled by provisional courts. With the end of the Second World War, the establishment of a permanent international criminal court became an agenda item of the UN. In 1948 the International Law Commission (ILC), the UN body assigned to develop code of international law, developed two aspects of international criminal: the drafting of a code of crimes against humanity, and the drafting of statute for the establishment of an international

---

81      De Temmerman, E., p. 10.

criminal court. The four crimes over which the ICC has jurisdiction are listed in the Rome Statute to include: genocide, crimes against humanity, war crimes, and aggression.

The LRA and its Commander Joseph Kony notoriously adopted the strategy of instilling fear to intimidate civilians to either join them or stop them from collaborating with the UPDF. In pursuit to this strategy the LRA resorted to tactic of cutting off the lips of people they accused to be collaborating with the government. This strategy was also serve as a punishment to those among the Acholi people who failed to register their allegiance to the LRA. Margret Gacega explained the LRA atrocities in the Acholi land as below:

> In northern Uganda, the Lord's Resistance Army (LRA), one of the world's most treacherous guerrilla forces operating under the leadership of John Kony, has been abducting children to serve as soldiers.... Female children are used as sex slaves.... The phenomenon of trafficking women and children has exposed them to sexual exploitation....[82]

## UPDF Human Rights Abuses

There is another side of the coin. The Ugandan army has also been accused of atrocities including the rape and killing of civilians, and many Northerners blame the Ugandan government for forcing them into inhumane camps and for failing to protect them from the marauding LRA[83]. In 2005, another Human Rights Watch reported that the Ugandan military were continuing to kill, rape and uproot

---

82      GECAGA, M. The Plight of the African Child: Reflections on the Response of the Church. Studies in World Christianity, p.14.

83      Worden, S. (2008). The justice dilemma in Uganda. United States Institute of Peace.

citizens in Northern Uganda[84]. They have also reported that a particular contingent of UPDF personnel on protection missions committed numerous deliberate killings and beatings of civilians in early 2005, when they were assigned to the displaced persons camps[85].

In 2007, Save the Children reported further abuses committed by the UPDF[86]. In its report, it mentioned that the UPDF was responsible for the deaths of nearly 100 children in an incident in the Karamoja region. These allegations were confirmed by a report from the UN, which urged the Ugandan government to curb human rights abuses against civilians and condemned 'indiscriminate and excessive use of force' by the Ugandan military.

Similar human rights abuses by the UPDF were also reported in other countries where UPDF forces were deployed on peacekeeping missions. In 2007, UPDF soldiers were sent to nearby Somalia to participate in the UN-supported African Union Mission in Somalia (AMISOM), the purpose of which was to provide peacekeeping services during the Somali Civil War. While in Somalia, UPDF soldiers reportedly sexually abused and exploited vulnerable Somali women and children at their bases in Mogadishu. Further, a Human Rights Watch report found UPDF soldiers raped or assaulted women who came to their bases and paid vulnerable women for sex. This directly contravenes the UN Secretary General's rules which prohibit peacekeepers from exchanging any money, goods or services for sex. As a

84      http://www.hrw.org

85      ibid

86      War Child UK, September 2010. A study of the community-based Child protection mechanism in Uganda and the Democratic Republic of Congo. https://resourcecentre.savethechildren.net/node/3021/pdf/3021.pdf

result of this, some women were found to have contracted sexually transmitted infections after the assaults, with several also describing being slapped and beaten by the soldiers.

Similar reports were echoed in an investigation which uncovered evidence of UPDF sexual exploitation of women seeking medicine for sick babies at AMISOM military bases and reports that UPDF soldiers gave some women food or money in exchange for sex, which the report suggested that it was an attempt to frame the assault as transactional sex. However, the UPDF suggested that only one rape case was ever brought to Ugandan military court, which saw a number of soldiers from UPDF suspended for misconduct. In an official response, AMISOM said that the alleged rapes did not reflect their soldiers' conduct and were considered isolated cases.

In 2009, members of the UPDF were sent to the Central African Republic to suppress the LRA's activities there. The UPDF presence was further increased in 2011 and 2012 in an attempt by the African Union to eliminate the LRA. In 2016, the UN reported a high number of rape cases by the UPDF in the Central African Republic, including cases involving children. In response, the UN High Commissioner stated that he was deeply concerned by these "credible and deeply worrying" allegations of human rights violations. Human Rights Watch found accounts of rape and sexual exploitation by the UPDF in interviews that they conducted with women in the country. Similar accounts were revealed in a BBC report, which detailed how a twelve-year-old girl was raped by a UPDF soldier on her way to the market.

Another account suggests that there have been a number of accounts of CAR women left pregnant by Ugandan soldiers when they returned to their home country. Reports suggest that women have had sex with UPDF soldiers in military bases despite strict rules explicitly prohibiting this. In each case, the alleged soldier who had fathered the child subsequently left the country and provided no support to the mother. Ugandan military investigators have claimed to have engaged with some of those affected, but this has been widely repudiated by survivors. One young woman told Human Rights Watch that she was warned not to speak with Ugandan investigators. A military spokes-person quashed the allegations, stating that the investigations were complete and had found such claims false.

In 2017, the UPDF withdrew from the Central African Republic. However, the Ugandan Government has still failed to address the se-rious, credible and consistent accounts of sexual abuse and violence committed by Ugandan soldiers in Uganda, Somalia and the Central African Republic. Critics argue that investigations by the Ugandan military have been wholly inadequate and have not held alleged perpe-trators to account. Some human rights organisations have now called on the African Union and the UN to undertake independent inquiries into the actions of the soldiers and to require the Ugandan govern-ment to act if the allegations are corroborated.

**Internal and International Response to the Human Rights Abuses**
As the LRA-Uganda government war intensifies with serious human rights violations being recorded on both sides, the Uganda govern-ment and the international community became concerned about these violations. However, much of the blame for these violations was point-ed at the LRA. As Worden elaborates:

*Extreme forms of brutality have characterized the LRA's insurgency. In the past 15 years, the group has kidnapped tens of thousands to be used as fighters, laborers, and concubines. In the process, thousands of others have been killed in attacks, or maimed to set an example for other civilians who would consider resisting. This violence, in turn, produced up to 1.8 million IDPs who were gathered in refugee camps throughout Northern Uganda.[87]*

The other move by the international community was to respond to the Ugandan Government's call, which asked the International Criminal Court to intervene by investigating the LRA's atrocities in northern Uganda and South Sudan. This move was intended to make the LRA's conflict set a precedent in the international justice system.[88] In January 2004, President Museveni's government's call to investigate the LRA's activities in northern Uganda was heard and resulted in an agreement with the ICC to investigate LRA activities in northern Uganda.[89] The Government of Uganda's main demand and objective was to let ICC prosecute Joseph Kony and his commanders and to denounce the LRA as a designated terrorist organisation.[90] Joseph Kony and his three top commanders were eventually charged with crimes

---

87      Worden, S. (2008). The justice dilemma in Uganda. United States Institute of Peace.

88      Mareike Schomersus (2007). The Lord's Resistance Army in Sudan: A History and Overview. Small Arms Survey. P. 10.

89      Sherif M. Bassiouni (1996). 'Searching for Peace and Achieving Justice: The Need for Accountability'. Law and Contemporary Problems. 59 (4). P. 9-28.

90      Heike Behrend (1991). 'Is Alice Lakwena a Witch? The Holy Spirit Movement and its fight against Evil in the North'. In Holger Bernt Hansen and Michael Twadelle. Eds. Changing Uganda: The Dilemmas of Structural Adjustment and Revolutionary Change. James Currey. P. 162-77.

against humanity, subsequently leading to their indictment.

The LRA's central strategy is the abduction of civilians, mostly children. The abducted are used to carry items looted from raided villages; most are taken to the LRA bases to be trained as combatants and deployed to the frontline, while some others, especially the girls, become sexual slaves and/or domestic workers. The abductees are tortured or killed if they attempt to escape. The World Development Report[91] of 2007 estimates that 66,000 children have been abducted by the LRA since the conflict began in 1986. In 2003, because of the increasing number of abductions, tens of thousands of children in Northern Uganda, commonly referred to as 'night commuters', travelled miles on foot to towns and city centres to sleep in bus stations, churches, storefronts and on the street.

By 2005, at least 50,000 children had made this nightly sojourn. Parents who were afraid that the LRA would attack or abduct their children if they stayed in the villages or IDP camps overnight sent them to the towns. The children, particularly girls, were at risk of sexual violence and other forms of abuse; they were attacked on their way to or at their sleeping places, as there was no protection offered during the journey and no supervision in the night. In April 1989, the government ordered people out of their homes and into 'protected villages' or IDP camps, which worsened the crisis. The government forced nearly two million people from their homes to IDP camps, and entire villages and gardens were razed to the ground to cut off means of subsistence and ensure that civilians left. The IDP camps

---

91      Fares, J., Gauri, V., Jimenez, E.Y., Lundberg, M.K., McKenzie, D., Murthi, M., Ridao-Cano, C. and Sinha, N., 2007. World development report 2007: development and the next generation (No. 35999, pp. 1-378). The World Bank.

were overcrowded, and they lacked basic social services like education, health, water and sanitation. Surveys and reports estimated that about 1,000 civilians died per week in the camps from malnutrition, poor sanitation and fires that often ravaged the camps. In addition, the confinement of people in the camps without adequate protection made it easier for the LRA to carry out raids and massacres.

One of the most notorious massacres perpetrated by the LRA was carried out at Patongo, Pader district in November 2002, where the LRA murdered twenty people and the commander of the group ordered that two bodies be dismembered and boiled in a pot in the presence of survivors. The first massacre perpetrated by the LRA took place in April 1995 in Atyak, when a team of LRA rebels commanded by Vincent Otti, who was born in Atyak, lined up more than 200 people on the bank of a river and shot them in cold blood. In July 1996, the LRA killed at least 150 Sudanese refugees in a succession of attacks in Acholi camp, a Sudanese refugee camp in Northern Uganda. In January 1997, the LRA clubbed or hacked to death at least 400 civilians in villages in Lamwo County. In July 2002, the LRA killed 90 civilians, most of them children in Pajong Village, Mucwini, Kitgum district. In October 2002, the LRA killed at least 120 civilians in Amel Village. In February 2004, the LRA killed at least 300 civilians, most burned to death, in Barlonyo IDP camp in Lira district. And there were many other massacres.

Criticisms pointed at the UPDF on how it handled civilian protection against the LRA attacks were not uncommon. It is observed that massacres perpetuated by the LRA in northern Uganda occurred near UPDF locations, especially during 1995. It was only after mass LRA atrocities had been committed and the LRA had departed that the UPDF would arrive. Further, issues of human rights abuses and harassment of civilians during UPDF operations have been recorded. Adam Branch writes:

> *The Ugandan government's counterinsurgency has also been brutal toward Acholi, as the NRA and its successor, the Uganda People's Defense Force (UPDF), have focused their use of force on destroying suspected rebel support among civilian[92]. Indeed, the Government soldiers perpetuated violence increased during Operation North in 1991. The NRA soldiers as alleged by the Acholi themselves carried out a number of massacre and other atrocities to deter those who sympathized with LRA. It was noted another period of intense government violence was seen in Gulu district especially in September 1996, when the government instituted and introduced its policy of protected camps. This policy became a forced displacement and drove hundreds of thousands of Acholi rural population out of their villages into camps. The NRA effected this policy through a campaign of disappearance of government critics, intimidation, and more seriously through indiscriminate air and artillery bombing and intentional burning of villages to scare out the residences. This was followed by intimidating military repeated ultimatums with deadlines that anyone who fails to relocate to the designated camps would be considered a LRA rebel or its sympathiser which put his/her life at risk with high chance of being killed.*

The critics of this government policy of forced displacement argued that while the government euphemistically calls the camps 'protected

---

92      Branch, A. (2007). Uganda's civil war and the politics of ICC intervention. *Ethics & International Affairs, 21*(2), 179-198.

villages', they are in other view considered as internment or concentration camps. This is so believed because of the nature the implementation mechanisms were so forceful rather than being voluntary. The government soldiers used brutal force to induce continued forced displacement in order to keep civilians submit to the idea of protected camps and see them as for their own safety. The civilians were left with no choice but to camp in the so-called protected camps. By the end of 1996 the total population in these camps have reached hundreds of thousands. Few years later the population had grown to a million, almost nearly the entire rural population of the Acholi sub region. Thus, the level of government soldiers violence against Achloi civilians during military campaigns such as Operation North, 1992 was coupled with murder, rape, the enlistment of children, arbitrary arrest, massacres and other atrocities[93]. However, this can never be compared with LRA atrocities but it rose an eyebrow both internally and internationally.

While the UPDF obviously has not been spared from allegations of committing human rights violations in northern Uganda, other voices have also connected these violations were due to incompetency and the lack of military professionalism in the UPDF government forces. This very view has been echoed by Robert L Feldman. According to Feldman the UPDF shortcomings were numerous, beginning with a general lack of competent to dedicated military professionalism"[94].

Indeed, in 2003, the Human Rights Watch released a report about the government abuses during the LRA war. This report

93      Branch, A. (2007). Uganda's civil war and the politics of ICC intervention. *Ethics & International Affairs, 21*(2), 179-198.

94      Feldman, R.L., 2008. Why Uganda has failed to defeat the Lord's Resistance Army. *Defence & Security Analysis, 24*(1), pp.45-52.

documented human rights abuses committed against the civilian population during the conflict against the LRA by the UPDF. These abuses included child soldiers being recruited into the army against their will and sometimes subjected to torture[95]. The report alleged that girls as young as twelve had been raped and had subsequently tested positive for HIV, most likely as a result of the sexual abuse[96].

## Conclusion

Following NRM/A and Museveni's control of Kampala, the capital city of Uganda, in 1986, General Tito Okello's defeated army and civilian supporters fled army barracks and cities and grouped and formed the Uganda People's Democratic Army (UPDA) that led the first revolt in Northern Uganda. It did not take long for the NRA to defeat the UPDA, marking an end to UPDA activities in northern Uganda in 1987. After the defeat, some of the fighters received amnesty in accordance with the Amnesty Statute of 1987, which provided general and specific pardons to groups that were engaged in rebellion.[97] The defeat of UPDA marked a formation of another rebel movement, mainly from the remnants and other dissidents who refused to accept amnesty and came together to form the HSM.[98] Within a year, Lakwena gained

---

95      Will Ross, Uganda army in 'rights abuses. BBC, Kampala, Uganda 16 July 2003.

96      Will Ross, Uganda army in 'rights abuses. BBC, Kampala, Uganda 16 July 2003.

97      See Prudence Acirokop. Accountability for Mass Atrocities the LRA conflict in Uganda. LLD Dissertation, University of Pretoria August 2012, p. 2.

98      See Allen & Vlassenroot (n 4 above) 8; see also 'Northern Uganda Chronology' Conciliation Resources http://www.c-r.org/our-work/accord/northern-uganda/chronology.php.

and maintained the HSM insurgency, reaching within eighty kilometres of Kampala before being defeated and fleeing into exile.

In 1988, Joseph Kony, himself a relative of Lakwena, claimed that the Holy Spirit had anointed him to continue and complete the work that Lakwena had started but left unfinished.[99] Indeed, he took over the command of the movement and by 1994, Kony's small group of rebels came to call itself the Lord's Resistance Army.[100] Like Lakwena, Kony promised to overthrow President Museveni's government and to purify the Acholi people from within. He claimed that the Holy Spirit had revealed that both goals would be accomplished through violence. Thus, in fulfilment of this vision, Kony's early campaign, though not particularly significant, maintained a degree of insecurity in northern Uganda.

This changed in 1994, when Sudan began supporting the LRA in retaliation for the alleged support that President Museveni's government was giving to the Sudan People's Liberation Army (SPLA). The LRA from then on got a sound base in South Sudan, where they received training and weaponry, stepping up operations in Northern Uganda, where they mainly attacked civilians.

In Uganda, ethnic conflicts have been pervasive since independence from the British in 1962. Since 1986, Uganda has been ruled by the National Resistance Movement (NRM) led by Yoweri Museveni, whose main constituency has been viewed as the Bantu-dominated South. Based on this notion of tribalism and regionalism, the Museveni government has, since ascending to power, faced opposition and armed

---

99      See Profile: Joseph Kony' BBC News, 7 Oct 2005 mhttp://news.bbc.co.uk/2/hi/africa/4320858.stm; Allen & Vlassenroot, and Gersony.

100     'Northern Uganda: The Forgotten War' Catholic Relief Service 2004 http://icar.gmu.edu/ICC/NorthernUganda.pdf.

rebellion in several parts of the country, especially in northern Uganda, where the LRA was active until recently, and close to the border with the Democratic Republic of Congo, where the insurgency led by the Allied Democratic Forces (ADF) was alive until 2004.

Uganda's history since independence in 1962 has been dominated by a series of military coups and brutal regimes that were responsible for grave and systematic human rights violations.[101] The first post-colonial president of Uganda was the King of Buganda, Sir Edward Mutesa II. Milton Obote, from Northern Uganda and long-time opponent of autonomy for kingdoms in Southern Uganda, including the kingdom of Buganda, was prime minister. On 24 May 1966, Obote ousted Mutesa and ascended to the Presidency. Obote further suspended the 1962 constitution, abolished kingdoms and consolidated his control over the military by eliminating several rivals. The first regime of Obote ended in 1971, when his army commander, General Idi Amin Dada, took over power in a coup. Amin lasted for eight years, and after his fall in 1979, Professors Yusuf Lule and Godfrey Lukongwa Binaisa had short tenures. Uganda held elections in 1980 and although disputed, the election ushered in the second rule of Milton Obote. In July 1985, Obote's army general, Tito Okello Lutwa, overthrew Obote and assumed his position in government. Lutwa's rule ended in January 1986, when Yoweri Kaguta Museveni, took over power following his successful six years of irregular war. During these various regimes, the rule of law was suspended, and a series of crimes occurred, the LRA operating as a major abuser.

---

101    For example, Field Marshall Iddi Amin Dada has been known for that.

# Chapter 3

∘∞∞∞∞∞∞∞∞∞∞∞∞∞∞∞∞∘

# The Peaceful Approach

## Introduction

The Ugandan government has pursued a dual approach of military action and mediation to bring a meaningful conclusion to the LRA rebellion in northern Uganda. One of these approaches was the peaceful approach. The military approach, which President Museveni favored at the beginning of the LRA war seemed not to be the best way to conclude the rebellion in northern Uganda. After achieving a significant victory over the LRA, President Museveni then accepted peaceful settlement to the conflict. This should never surprise anyone because all conflicts are ended through peaceful means.

It is understandable why Museveni opted to the use of force against the LRA. Since its inception, the LRA's strategy has been focused on the forceful recruitment and conscription supplemented by a massive policy of abduction of under aged girls and boys. The LRA abductees were used to do all sorts of unpleasant jobs[102]. The LRA also employed tactics of terror on those who do not offer themselves to join its rank and file, including mutilating body parts. Apart from making people submissive to the LRA, this policy served another purpose. It was intended to reduce the image of the NRM/A government and made it look incapable of protecting the citizens of Uganda.

In order to protect its image and to show the world that it was an effective and responsible government, President Museveni and its military the NRA resorted to military approached by choosing to battle the Joseph Kony's LRA insurgency in northern Uganda, which to the large extend succeeded in pushing the LRA away from the Uganda borders. This was quite an achievement for the Government of Uganda. The government justified its choice for military action rather than peaceful approach against the LRA on the basis that LRA and its leaders do not understand any language other than the barrel of a gun, and that it was through this means that LRA's abuses could be quickly addressed. Indeed, by 1990s the LRA war has already been described as one of the worst humanitarian crises in the world.[103]

However, having failed to achieve its objective to end the conflict militarily, the Government of Uganda as a second option, invited the International Criminal Court to investigate the activities of the LRA with an aim to prosecute its leaders, who are accused to be responsible

---

102    This include being porters of looted items to the LRA camps. Girls are compulsory married.

103    Kasaija Phillip Apuuli, The ICC Arrest Warrants for the Lord's Resistance Army Leaders and Peace Prospects for Northern Uganda, *Journal of International Criminal Justice*, Volume 4, Issue 1, March 2006, Pages 179–187, https://doi.org/10.1093/jicj/mqi092

for the most ever seen human rights abuses in the twentieth century. The LRA perpetrated grave abuses in this conflict.[104] What is interesting though with the LRA conflict is that it used multiple approaches all intended to arrest the conflict very quickly, ranging from military action by the Government of Uganda to the involvement of the international community through the International Criminal Court. This chapter examines various approaches used to address the LRA problem.

## The Acholi Initiative

Despite Museveni's military approach to ending LRA abuses, sizable internal and external initiatives to bring an end to the LRA war were also attempted, especially by the Acholi people themselves. As already mentioned, there was the 1993/94 peace dialogue led by Betty Bigombe, who was Minister of State in the Office of the Prime Minister of Uganda, resident in Northern Uganda. It is believed that Bigombe's post was created by President Museveni as part of his move to address the rebellion in Acholi land.[105] In reality however, Museveni created this post to counter criticism labelled against him that he himself was an obstacle himself to ending the LRA rebellion through peaceful means, and Bigombe was one of those promoting peaceful solution to the LRA rebellion. As mentioned earlier, Bigombe launched her own initiative to dialogue with the LRA combatants. It was based on her initiative, that Museveni created the position for her; to show that he

---

104      Ibid.

105      Maphosa ,S. B.. The Lord's Resistance Army: A Review of African Union Regional Efforts to Eliminate the Resistance in Central Africa in Festus B Aboagye Ed. The Comprehensive Review of the African Conflicts and Regional Interventions, 2016. P.196.

was indeed genuine for ending LRA war by all means and that he was supportive for the peaceful settlement. More importantly, the president established a Presidential Peace Team (PPT), headed by Bigombe, to carry on with talks with LRA rebels. Unfortunately, by November 2002, the LRA had shown its lack of interest in peace and ignored a call from the newly established Presidential Peace Team (PPT), which called for members of the LRA to assemble in designated 'safe-zones'. There was a problem with PPT. They seemed to have jumped the gun. The process to assemble combatants in designated safe zones always comes when talks are in the final stages and when confidence building has been achieved. It was thus premature to call for assembling at this stage. The LRA interpreted this move as trap.

Indeed, the LRA refused Museveni's appeal to assemble in safe zones and instead demanded that a ceasefire be extended throughout the whole region[106]. The demand that was unacceptable to President Museveni. Prior to this, the government had earlier declared a seven-day ceasefire, to enable Betty Bigombe pursue talks with the LRA - overseen by the US, the UK and the Netherlands.[107] Although viewed elusive, Betty Bigombe's initiative was remarkable. It became the first ever attempt by the Government of Uganda to address the LRA problem by dealing and establishing direct contacts with LRA leadership through the Acholi people. Perhaps Bigombe had learned something from the previous local involvement, where community voice was critical in finding an end to the conflict. As Mwaniki et al writes:

106    Dolan, C. (2010). Peace and conflict in northern Uganda 2002-06. Accord-an international review of peace initiatives., Update to Issue, 11, 8-9.
107    Ibid.

*While it can be said that the 1993/94 Bigombe peace talks were the*
*first real attempt at negotiations with the LRA, the 1988 government*
*armistice with the Uganda People's Democratic Army (UPDA) rebels*
*saw the communities requesting the inclusion of all other rebel outfits*
*in the peace process. The community voice was critical, and so were the*
*efforts by the diaspora.*[108]

By involving Acholi politicians and various religious denomina-tions in the Acholi region of which Mrs. Bigombe was one, Museveni washed his hands of it and threw the ball to the Acholi community themselves, while watching whether the Acholi leaders who spoke so confidently could bring peace. Museveni had hoped that the Acholi-based mediations would build confidence and trust in the government which the Acholi people had viewed until then, to be their enemy. Unfortunately, the Acholi initiative collapsed before it reached any-where, citing lack of political will from the side of the government. This point is illustrated by O'Kadameri as he writes:

*It seemed Betty Bigombe was sent to Acholi not to negotiate peace, but*
*to convince remnants of the insurgents to come out of the bush. For*
*five years most of what she did was to encourage the locals to tell their*
*sons to give up the rebellion. The decision to start talking peace was a*
*personal one by Bigombe, not backed by any official policy to end the*
*war through dialogue.*[109]

---

108    Mwaniki, D., Wepundi, M., & Morolong, H. (2009). The (Northern) Ugan-da Peace Process: An Update on Recent Developments.

109    O'Kadameri B. LRA/Government negotiations 1993-1994, p.35

Encouraging combatants to abandon rebellion cannot be interpreted as a lack of political will because there no better strategy than appealing to the combatants to abandon rebellion and accept peace. It was therefore duty of the LRA negotiators to bargain at the table as others do. The question of whether Museveni had been against peace did not arise after he had appointed a peace team.

Following Betty Bigombe's failed attempt to broker peace between the government and the LRA, the Government of Uganda ventured into another effort meant to address the LRA conflict. Thus, in 2000, there was second government attempt. As part of this initiative, the Government of Uganda enacted the Uganda Amnesty Act, which encouraged rebels to return home without fear of prosecution for acts of violence committed during combat against the government. Other objectives provided amnesty to rebels who renounced rebellion and gave up their arms; facilitated an institutionalised resettlement and repatriation process; provided reintegration support, including skills training for ex-combatants; and promoted reconciliation. The act also established an Amnesty Commission tasked with implementing the act and issuing certificates of amnesty. A report by the Enough Project suggests that a total of 12,971 former combatants from the LRA responded to this call[110]. Subsequent attempts by General Salim Saleh, (younger brother of President Museveni and who commended the northern front operations against the LRA) to enter into dialogue with the LRA in 2003 were also initiated. Unfortunately, this call was turned down by the LRA, on the basis that Saleh initiative did not

---

110      Conciliation Resources and Quaker Peace, "Coming Home – Understanding why commanders of the Lord's Resistance Army choose to return to a civilian life,"2006. Available at, http://www.c-r.org/sites/c-r.org/files/ComingHome_200605_ENG.pdf

cater for security and safety arrangements for LRA fighters.

After the collapse of efforts to bring peace by negotiation, the Government of Uganda launched a military campaign (Operation Iron Fist). The immediate effect of the Operation Iron Fist was the ignition of intense and violent attacks in retaliation by the LRA. As a result, several thousand families, especially in northern Uganda, were uprooted from their villages. The Iron Fist operations managed to inflict losses on the LRA but failed to end LRA hostilities on the civil population. The international outcry against LRA atrocities and the call for the government to do more to stop these atrocities, compelled the government invite external intervention, by referring the LRA abuses to the International Criminal Court. Accordingly, in 2005, the International Criminal Court issued warrants for the arrest of the LRA's top leaders, namely Joseph Kony, Vincent Otti, Okot Odhiambo, Dominic Ongwen and Raska Lukwiya.

When the efforts of the Uganda government are evaluated, it is argued that its initiatives with regards to addressing the LRA rebellion between 1986-2005 ended unsuccessfully and thus, it is considered a great failure in the field of conflict resolution. The failure is attributed mainly to lack of trust and confidence building between the Government of Uganda and the LRA command. The trust could not be built between the government and the LRA due to the fact that neither President Museveni nor Joseph Kony were genuine in finding a peaceful solution. They both ignored that making peace ultimately requires a firm commitment from the leadership of the warring parties' and a high level of courage in order for the peace to be brokered. This was the case with the Government of Sudan and the SPLM/A in 2005. To ensure a permanent end to violence requires full determination and imagination of all parties to seek solutions and build trust between their combatants that have been in the heart of conflict.[111]

## The Government of Uganda's Initiative

Although Uganda's government was not been eager to engage in peace talks with the LRA, it has done so. Perhaps it realized that no military solution can bring lasting peace and stability to the country. However, due to lack of political will, not much was achieved, and the peace talk attempts eventually failed. Subsequently, between 1995 and 2000, the government launched Operation North as part of its second phase campaign to wipe out the LRA. The immediate effect of this move was the ending of the then Hon minister for Northern Uganda Betty Bigombe's initiated talks with the LRA. According to those interested in and watching Ugandan politics whom the author had a chance to talk to, Museveni's resort to military option was a planned policy which the NRM/A pursued since it came into power in Uganda and this was not surprising. Indeed, the NRM/A government had already made it clear from the start of the LRA rebellion, that it will pursue a military solution to end the LRA rebellion. The policy generated mixed opinions among Ugandans, with some arguing that Museveni's aggressive approach allowed little room for political solutions. In particular, the Acholi politicians who favoured a political solution to the conflict, criticised Museveni's approach of defeating LRA militarily. Traditional Acholi leaders have also strongly advocated the use of traditional conflict resolution and reconciliation ceremonies as mechanisms for

---

111     Address by former UN Secretary-General, Kofi Annan, in October 2011 to the Basque Summit in Spain, organized by Conciliation Resources.

reintegration in the post-conflict context.[112]

## The Government of Sudan's Initiative

The terrorist activities towards the end of 1990s compelled the US to adopt new strategy in foreign policy decision making towards North Africa and the Horn of Africa regions. This time the US dedicated it's time to see to it that some security prevails in the Horn of Africa. Its first move was to bring normalcy in relations between Sudan and Uganda. Following the US's facilitation of normalisation of diplomatic relations between Uganda and Sudan, the Government of Sudan authorised its First Vice President (and President of the Government of Southern Sudan) Salva Kiir Mayardit, to mediate a peace agreement between the Government of Uganda and the LRA. Salva Kiir Mayardit on his part appointed and tasked his deputy, Dr Riek Machar, to mediate a peace deal between the Government of Uganda and the LRA, thus, the Juba Peace Initiative was born. Unfortunately, this initiative did was not fruitful. This was because each of the three countries with interest in the LRA: the US, Uganda and Sudan, continued to pursue their own national interests and agenda, which were pursued bilaterally or multilaterally.

The US, being a world power, major aid donor and development partner in Sudan and Uganda, had an upper hand and veto powers to bring Sudan and Uganda together to the table after long sour diplomatic relations. The US had the power to use any means necessary—political, financial or military—to control Sudan and Uganda. Pedro Banos explains such global relations in the words below:

---

112     Rose, C. and Ssekandi, F.M., 2007. The Pursuit of Transitional Justice and African Traditional Values: A Clash of Civilizations-The Case of Uganda. *SUR-Int'l J. on Hum Rts.*, 7, p.101.

*On the global stage, there are essentially two types of country: the dominant and the dominated. The former exerts control on a regional or global scale, while the latter is controlled, more or less directly, and in various ways—militarily, economically, culturally or technologically...* [113]

Indeed, the US's influence on the Horn of Africa region in general and Sudan and Uganda in particular, is remarkable. With Sudan, the US's influence is deeply rooted in the geostrategic location of Sudan and Uganda. Sudan, being at the centre of African, Arab, Christian and Islamic politics, has held great interest for the US since its independence in the beginning of the 1956. Uganda, under Museveni, has been a reliable partner for the US in promoting stability in the Horn and Central Africa regions and in combatting terror. Therefore, the direct influence of US on these two nations goes a long way, though sometimes smoothly and sometimes with upheavals.

## The Government of South Sudan Initiative

The South Sudan Government's effort to mediate peace talk between the LRA and Uganda Government was led by the then Vice President Dr Riek Machar Teny. The location of the talks was Juba, the capital of South Sudan. The Machar chaired LRA Juba peace talk was the first South Sudan Government's attempt to bring peaceful resolution to the Uganda conflict and to end the LRA's atrocities in both South Sudan and Uganda. The LRA, for security and logistical reasons, crossed into South Sudan as early as the 1990s and stationed in Eastern Equatoria,

---

113     Banos, P. (2017). *How They Rule The World: The 22 Secret Strategies of Global Power*, London, Ebury press, p.4.

where it launched attacks on Uganda. It was easy for the LRA to be stationed in South Sudan and Eastern Equatoria because of historical and ethnic links with South Sudanese tribes on the border namely the Acholi and Lango tribes. The shared ethnicity and language commonality between the LRA combatants and South Sudanese tribes at the border, facilitated the LRA's quick and easy integration with the locals. Mareike wrote:

> One chief remembers that in the early days, people did not make the connection between the armed fighters who appeared in their region and the LRA...people started to recognize the LRA only when the fighters employed tactics familiar to Uganda and when they began abducting people.[114]

Without the recklessness of the LRA it would be impossible to distinguish from the locals, of the location they were based in South Sudan due to the fact that they speak the same language, look alike and have other common features. The difficulty in recognizing the LRA was indeed attributed to the tribal commonalities between the tribes along the common borders of the two countries. If the LRA had maintained discipline, the war against them would impossible, especially in South Sudan where it would be possible for them to melt into the local population.

Indeed, before the establishment of the Government of South Sudan (GOSS) as necessitated by the CPA, the LRA managed to recruit abducted South Sudanese youths into its forces, along with others from the Democratic Republic of Congo and the Central African

---

114     Mareike Schomerus. P. 18.

Republic. Those who were recruited became an integral part of the LRA movement, which became a source of their survival in South Sudan.

The LRA, through the influence of Sudan government, attacked SPLA bases and carried several deadly ambushes along the Magwi, and Nimule corridor. These attacks and ambushes continued till the formation of the GOSS. In 2006, the Government of Southern Sudan tried to bring the two warring Ugandan parties into negotiation[115]. This led to the start of mediation in Juba.

## The Juba Peace Talks

The Juba peace process which was based in Juba Raha Hotel, mediated by then Vice President of Government of Southern Sudan Riek Machar, was supported by the UN, whose envoy was the former President of Mozambique Joaquim Chissano. The involvement of UN as exemplified by the assignment of Chissano, showed no doubt that the world leaders were indeed in search of peace in Uganda. This involvement nearly achieved peace in March 2008 between the Government of Uganda and the LRA. Unfortunately, the nearly reached agreement abruptly collapsed few weeks before it was finally signed, with much blame being pointed on Joseph Kony for the destruction of the expected peace deal. However, there were other opinions that suggested lack of political will from the Uganda Government. As the International Crisis Group stated in their periodical assessment of Operation Lightning Thunder, 'this lack of commitment appears free of political consequences for Museveni, who is under little domestic pressure to

---

115     By then South Sudan was still under Sudan. South Sudan became independence on 9 July 2011

finish off the LRA'.[116]

## The Short-Lived Juba Peace Talks

Why the LRA Uganda Juba Peace Talks failed remains an elusive question? The rumour that the government of Uganda had no interest or political will in concluding peace with LRA spread like a wild fire, after Joseph Kony refused to sign the final deal. Likewise, Joseph Kony's refusal to sign the final deal was quickly interpreted to be a lack of commitment from Kony's side. However, neither the information could not be verified or ascertained as there was no senior delegation from the government side nor was Joseph Kony available to defend decision.

But all in all, some voices suggested that it failed for a number of reasons. Firstly, it was alleged that the parties arrived at the negotiation table for different reasons and with different agendas. Of course, this always happens and it is not uncommon for the parties to come to the negotiation's venue with different agendas. It is during the negotiations that positions may shift for the purpose of bargaining. It is further argued that each side was under international pressure and saw its involvement in the process as an opportunity to correct its image in the face of the international community. With the LRA squeezed as a result of the CPA, the LRA leadership saw the opportunity to present itself as a peacemaker working towards peace in Uganda, and it is not the brutal LRA as portrayed by the Government of Uganda. In this way, it is seen that the LRA's intention to participate in the peace talks was to avoid the growing international criticism. This could have been

---

116    The Lord's Resistance Army: End Game. Crisis Group Africa Report No. 182, 17 November 2011, p.3

one of the reasons, but there was another reason which compelled the LRA to attend the peace talks. The LRA had lost its logistics support from the Sudan because the SPLM/A, which was more allied to government of Uganda, had been handed over to the South Sudan administration based on the CPA. Upon taking over the South Sudan administration, President Salva Kiir told the LRA either you accept peace or leave South Sudan[117].

These contrasting agendas genuinely explained why the talks did not proceed. This contrast in expectations created a major obstacle to a successful resolution of the Uganda conflict. The government delegation offered an amnesty to Kony and his commanders on the condition that they accept peace and abandoned rebellion. Amnesty should have come as one of the carrots accompanied by other carrots. For example, sharing of power and resources. The sharing of power and resources did not feature during the discussions. This made Kony and his commanders question government sincerity. Additionally, this offer was parallel with military action against the LRA. Despite criticism, the government maintained its continued offensive was no contradiction and instead it argued that the continued military operations was necessary as part of pressure to push the LRA to make the it commit to peace talks. Sincerely, this later step contradicted spirit of peaceful solution and perhaps that's why the LRA did not commit to the deal with the government. On their part, the LRA delegation, who received full commitment of international community support for the talks, was dismissive of the amnesty offer and considered deadlines and military pressure as contradictory to the spirit of a political peace process.

117    Anonymous Interview, Juba 2016

The LRA, as already said, approached the negotiations with the government of Uganda as a chance to re-politicise the conflict that had come to represent seemingly senseless violence in the minds of world leaders. Getting the chance to express themselves in front of international press for the first time, the LRA delegation considered it a chance to redefine narratives on the war in Uganda and to identify what it saw as the root causes of the conflict. The LRA delegation argued that it had been forced to fight a legitimate war against an oppressive government of Museveni. They further argued that the inclusion of members of the Acholi diaspora in the LRA negotiations team should tell the mediators and the international community at large that the war they were fighting was indeed a justified one. Additionally, the presence of diaspora in the talks signalled that it was a time to address longstanding grievances among the people of Uganda. However, some observers during the peace talks believed that the LRA leadership's real motivation was to buy time to regroup and that its participation was primarily a tactical move.

A contentious issue from the above argument is the LRA commitment to the Juba peace talks. My take is that whether the LRA was sincere in its attempts to find a political settlement or not, the whole issue should be attached to the character of Joseph Kony. It is the character of Joseph Kony that must be seen a major element of the LRA's war strategy. The extensive press coverage that came with the LRA's participation in the Juba peace talks did not change the wide held perception about the LRA. The LRA damaged its own credibility as a political actor in Uganda and during the peace talks, as it continued to commit atrocities against civilians under the protection of the government that he was trying to bring peace with.

A second fundamental difficulty that featured during the Uganda Government-LRA Juba peace talks, was the about the nature of the skills of the delegates of the warring parties. The Uganda government's delegation was composed of nearly thirty members. Out of this number, twelve were active military officers mostly probably drawn from the military intelligence department. The rest were probably selected from other government branches of security organs, such as internal and external security directorates. In the delegation there were Museveni's most trusted high-profile politicians, such as Interior Minister Ruhakana Rugunda who led the delegation. In the delegation, also were Oryem Henry Okello, by then State minister for Foreign Affairs, Amb. Busho Ndinyenka and Hon Christine Aporu. There was no doubt the government delegation was well composed in terms of experience and skills but its composition dominated by the military officers made the LRA leadership and observers questioned the sincerity of the government towards the talks.

In contrast to the government delegation, the problem of representation was much more acute among the LRA delegation, which made the LRA draw its team composed mainly of members of the Acholi diaspora. This included David Nyekorach Matsanga, James Alfred Obita and Martin Ojul. This shortcoming was noted by the LRA members themselves. Members of the LRA delegation complained that they lacked technical assistance for research and were unable to present themselves as equal partners to the government delegation. Another shortcoming from the LRA delegation, was that they were not given full powers to negotiate because the LRA leadership did not believe in them, for the obvious reason that they were not real and committed members of the LRA. The LRA leadership feared that these diasporas will be pushing their own personal agendas. Indeed, as the peace talks progressed, it was observed that some LRA delegation members mainly from diaspora pushed through their own personal agendas and political interests, thus, causing tensions within the LRA delegation, particularly between the LRA bush members and the diaspora members[118].

This created a difficulty as they did not access the LRA high command. This shortcoming was realized by Dr Riek Machar, the chief mediator of the LRA peace talks. He tried unsuccessfully to strengthen the LRA delegation by convincing Joseph Kony and Vincent Otti, the LRA's second-in-command, to join the talks.

Thirdly, violation of cessation of hostilities was another blow to the Juba Peace first talks. There was constant violation by both sides to the cessation of hostilities agreement. This agreement was the first article to be signed by both warring parties and this was supposedly in good faith to allow confidence building between the delegations. The violation was inevitable because of the vagueness and lack of financial support to monitor the agreement on the cessation of hostilities. The assembly areas for the LRA forces were neither clearly demarcated nor agreed upon by the parties. The Government of Uganda proposed Ri Kwanga on the western side of the Nile. The LRA noted that the proposed eastern assembly area had been fully surrounded by Ugandan troops and thus, not safe. As discussions on the assembly area went on, different military sources on the ground reported UPDF helicopter attacks on the LRA. The LRA, on the other hand, retaliated violently by attacking Ugandan troops, civilians and South Sudanese civilians.

---

118     Interview with anonymous member of the LRA delegation, Juba May 26, 2008

Such violations damaged the credibility of the Juba talks, undermining expectations that peace negotiations would bring an end to violence.

Fourthly, the insistence of the International Criminal Court on their position to prosecute LRA leaders was another issue that made the Juba Peace Talks failed. Mixing justice issues as brought forward by International Criminal Court, which advanced arrest warrants for top LRA leaders, brought the relationship between peace and justice into question during the talks and divided local and international opinions. Many advocates for Joseph Kony and his commanders' trials by the International Criminal Court were facing a dilemma between backing the talks and protecting the image of the International Criminal Court. On the other hand, Joseph Kony and his supporters argued that relinquishing their indictment by the ICC was meaningful for the success of the negotiations and a gesture of good faith for political compromise with the LRA. The ICC on taking justice forward did not bring political incentives for final agreement.

Fifthly, operationally, the talks were proved to be a major challenge to the international community. After initial confusion of who should facilitate the peace process, the UN Office for the Coordination of Humanitarian Affairs (OCHA) took the lead, culminating in OCHA head Jan Egeland's visit to the LRA in the bushes of Southern Sudan and Eastern DRC in late 2006. The OCHA struggled with the task, and as Egeland explained, 'the fact that a humanitarian delivery agency like OCHA had to facilitate a political process at Juba exposed problems of coordination and capacity within the UN system'.[119] This brought about the establishment of the UN Juba Initiative Fund.

---

119    Jan Eagland, 'Uganda Conflict Worse than Iraq', interview with BBC News Channel, Monday, 10 November 2003.

The establishment of the UN Juba Initiative Fund (JIF) in October 2006 initially appeared to address the talks' early financial problems, promising large cash injection. But this too introduced another new problem. The sudden influx of money contributed to a rift within the LRA delegation and their field commanders. In general, oversight of progress at the talks was insufficient to encourage the parties to overcome their mutual distrust and negotiate in good faith. In an attempt to rectify this shortfall in trust, former Mozambican president Joaquim Chissano was appointed as UN special envoy to LRA-affected areas in late 2006 and was credited with bringing international attraction to the process and maintaining its momentum, as his appointment seemed to emphasise the UN's commitment to treating the conflict as both regional and political. But he was also criticised for not taking a clear stance on the military offensive launched by the Ugandan Army in December 2008.

The sixth factor was the position of the US in Washington. Washington's position in relation to the conflict in Uganda was regarded as not straight forward, as it seemed to mix attempting a political solution with a new approach of strengthening local military capacity. A US representative joined the mediation team in 2007, almost at the same time as the establishment of AFRICOM was announced. Confusion over AFRICOM's mandate added to the suspicion that it would provide military support to the Uganda military to use this support against the LRA.

The lack of progress with the LRA negotiations strengthened the position of defeating the LRA by military means. This was a plan B for the US, Uganda and South Sudan. In fact, joint military action was envisaged during the peace talks in Juba as an alternative for ending the LRA war, especially with the collapse of the first option of attempting to use peaceful means. This option was favoured by the Uganda Government, which had earlier indicated that the LRA only understands the language of war. However, international pressure forced the Uganda Government to accept talks with the LRA, and the result was the Juba peace talks. Despite the international community's desire for peaceful settlement of the LRA conflict, the Juba peace talks called for military action against the LRA in case they defied peaceful solutions, but this opened room for criticism for not applying the same policy against the Uganda Government.

The failure of the Juba peace talks was mainly about the trust and lack of confidence-building between Uganda and the LRA leadership. Further, the warrant for the arrest by the ICC of LRA commanders contributed negatively to the talks. Moreover, Sudan was never genuine in providing accurate information about the LRA. Instead, they continued their collaboration. For example, the LRA Crisis Tracker in 2016 reported information received from LRA defectors that Sudanese armed forces stationed in South Darfur helped introduce LRA men to traders in the area.[120]

The cooperation signed on the 4 March 2002 gave the UPDF access to conduct military operations against the LRA inside Sudan. It also asked the SPLA and SAF to commit themselves to the operation against the LRA. Paragraph nine of the Gulu Communique clearly states that:

*...considering the continued suffering of the people of Uganda and Southern Sudan caused by the LRA, the SPLA and the UPDF shall*

---

120      Ibid.

*continue to protect the peoples of Southern Sudan and Northern Uganda until the LRA decommissions or/and signs a comprehensive peace agreement.*[121]

Such public announcements were meant to deter the LRA and compel it to have faith in the talks but unfortunately it was not. Before and in 2002, the UPDF had launched a military campaign against the LRA. This campaign was the Iron Fist Operation already discussed above. The Uganda Government and the LRA's negotiations team in Juba were officially opened in the assembly hall of South Sudanese parliament in order to give weight as was confirmed by enormous public attendance.[122] The Juba peace talk was the first meaningful attempt to solve the LRA problem politically.

## Conclusion

Several attempts to address LRA conflict in northern Uganda were first initiated by Uganda themselves. The Uganda initiatives included an initiative by the Acholi people themselves and also there were initiatives from the government itself. Externally, there was move from South Sudan aimed at facilitating peace talks between the government of Uganda and the LRA. All these initiatives-internal and external failed because of the lack of political will either from the side of the government or from the side of the LRA. The Juba Peace Talks failed because Joseph Kony the LRA leader declined to sign it at last minute due to lack of political assurance.

---

121    Gulu Communique minutes, May 1, 2007.

122    Simonse, S., Verkoren, W., & Junne, G. (2010). NGO Involvement in the Juba peace talks: the role and dilemmas of IKV PAx Christi. The Lord's Resistance Army: Myth and Reality, 223-242.

# Chapter 4

◦◦◦◦◦◦◦◦◦◦◦◦◦◦◦◦◦◦◦

# International Relations

## Introduction

This Chapter places LRA war in the international context, by looking at the US and Uganda diplomatic relations since independence. It also reviews in brief Uganda, US and Sudan relations since the independence of Uganda in 1962. It places the LRA into historical realities of these countries.

## US - Uganda Diplomatic Relations

The US established diplomatic relations with Uganda in 1962 following Uganda's achievement of independence, on 9 October 1962, from the United Kingdom (UK)[123] However, post-independence Uganda,

---

123    U.S. Relations with Uganda BILATERAL RELATIONS FACT SHEET, Bureau of African Affairs, 26 October, 2018 . This information is available at https://www.state.gov/u-s-relations-with-uganda/

became unpleasant, with the country enduring despotism and economic digression. Equally, human rights abuses increased sharply, with many distinguished Ugandans losing their lives, especially during Amin Dada. This abuse of rights of Ugandans strained US diplomatic relations with Uganda for several years, to be reviewed only following Museveni takeover of capital Kampala on 26 February 1986. With the success and coming of President Museveni into power, Uganda experienced relative political stability and economic growth which attracted international and regional development intervention.

The period beginning 1986 -2001, Uganda, under Museveni, was seen as one of the few African countries reliable partner of the US which promoted stability in the Horn, East and Central African regions by taking part in combatting terror, particularly through its contribution to the stability of Rwanda, the DRC, Sudan, and to the African Union Mission in Somalia. In appreciating Uganda's role in bringing peace to the region, the US in turn, provided significant development and security assistance estimated a worth total assistance budget exceeding $970 million annually. Part of this US Government assistance was used to professionalise the Uganda military; providing anti-retroviral treatment for more than 990,000 HIV-positive Ugandans; and working to boost economic growth and agricultural productivity, improve educational and health outcomes and support democratic governance through inclusive, accountable institutions. On economic revitalization, the US Mission (US Embassy) worked very closely with the Government of Uganda to improve tax collection, oil revenue management and to increase Uganda's domestic funding for public services and the national response to HIV/AIDS, among many others.

All these were part of the realization of the African Growth and

Opportunity Act (AGOA) a policy that the US designed to help African countries benefit from US trade. Uganda has been one of the few African countries selected for benefit under this program. Under this arrangement US exported to Uganda agricultural machinery, optical and medical instruments, wheat and aircraft in exchange for Ugandan coffee, cocoa, base metals and fish. To advance this relationship, the US committed to signing trade and investment framework agreements with the East African Community and with the Common Market for Eastern and Southern Africa. Uganda is a member of both regional organisations.

The US joined Uganda to fight the Lord's Resistance Army (LRA) in 2011, when President Obama succumbed to the pressure of US law makers such as Sens. John McCain and Russ Feingold, who sought US intervention to stop LRA atrocities in northern Uganda. The US government, going back to the time of Bill Clinton administration, has established a strong working relationship with Museveni, given its geo-strategic position to pressure Sudan and thus, a partner in combating international terrorism. Also, Museveni has supported and worked closely with the US in controlling HIV/AIDS. Until five years ago the US had supported military efforts in Uganda to end the LRA war, providing no impetus to catalyze a peaceful resolution of the conflict[124].

## US - Sudan Diplomatic Relations

In regards to the diplomatic relations between Sudan and US, the US established diplomatic relations with Sudan at the time of Sudan's decolonisation in 1956, following its independence from the joint Egypt

---

124    Quaranto, P. J. (2006). Ending the real nightmares of northern Uganda. *Peace Review: A Journal of Social Justice, 18*(1), 137-144

and UK administration. This quick diplomatic relation was promoted by common interest. The US was decolonized by the British in 1776 and since then, it had become a champion of advocacy of independence for other nations who were still under colonial rule. In other words, the US was happy to see that other countries were decolonized too. The US Sudan diplomatic relations saw a number of development projects supported by US, through its development agency USAID, being implemented in Sudan. South Sudan in particular benefited from US AID. South Sudan saw a number of schools which became a base of education built with US help.

In 1967, however, Sudan broke diplomatic relations with the US following the start of the Arab-Israeli War.[125] In 1972, after five years of no diplomatic contact between the nations, relations were re-established. The move was probably encouraged by the Addis Ababa Agreement, which ended seventeen years of civil war in Sudan between the south and the north. Despite improved relations, US–Sudan relations have not been stable because Sudan has linked itself to terrorist organisations, leading the US to accuse Sudan of establishing links with international terrorist organisations and eventually resulting in the US's designation of Sudan as a state sponsor of terrorism in 1993, and the suspension of US Embassy operations in 1996, with the US Embassy in Sudan only reopened in 2002. Following this diplomatic move, US policies in Sudan have focused on ensuring that Sudan does not provide support to, or a safe haven for, international terrorists; achieving a definitive end to gross human rights abuses; concluding a comprehensive peace process; and encouraging an open and inclusive political dialogue to address constraints on personal, political and

---

125    Department of State. US Relations with Sudan. 15 November 2015.

public expression.[126]

According to Johnson, initially, the Clinton administration had no policy on Sudan, but Clinton's inauguration in January 1993 was predicted to be a departure of vocal Republican figures, who had defined the region in the context of the Cold War.[127] Johnson argues that 'Khartoum's hostility to its neighbours became a factor in defining the US's attitudes towards its former ally. Association with other militant Islamists such as Hamas and Usama Bin Ladin and continuing military ties with Iraq and Iran were further reasons for the US condemnation of Sudan as a terrorist state'.[128]

## US – South Sudan Relations

Following the formation of the Government of South Sudan, as part of the CPA requirements, one of the priorities of South Sudan was promotion of the already existing diplomatic relations with the US. So, when the US demanded South Sudan to be involved in the fight against the LRA, it became a golden opportunity for South Sudan to repay the US and Uganda for their support for the South Sudan struggle. Both US and Uganda sacrificed a lot for South Sudan to achieve its independence. For this reason, it was very easy to bring South Sudan on board to fight the LRA, although this was not the first time for SPLM/A and Uganda's NRM/A to join hands against the LRA. They had fought many battles against the LRA, especially between 1995-2005. As Gerard Prunier highlights:

---

126    Ibid.

127    Johnson, D.H., 2016. The Root Causes of Sudan's Civil Wars: Old Wars & New Wars. James Currey. p. 102.

128    Ibid.

*"In many ways Sudan and Uganda have been running an undeclared war on their common border since 1986. Sudan has been supporting a bizarre syncretic and millenarian movement, the Lord's Resistance Army (LRA),[1] which is still fighting the Museveni regime in northern Uganda. Meanwhile, Kampala has progressively given increased help and facilities to the Sudanese People's Liberation Army (SPLA) which is fighting the Khartoum regime in the southern Sudan"[129].*

Behind this collaboration was the US. The United States has been the largest provider of bilateral foreign assistance to South Sudan and a major financial contributor to peacekeeping efforts there. The United States historically supported self-determination for the South Sudanese and played a major role in facilitating the 2005 peace deal, that brought an end to Africa's longest-running civil war. Congress was active in supporting South Sudan's independence and plays an ongoing role in setting US policy toward both South Sudan and Sudan. South Sudan recognizes the role played by the US to the extent that it intended to have its independence shared with the US but for practical reasons, it missed by five days. South Sudan's independence was on 9 July 2011. The US got its independence from British on 4 July 1776 - two hundred and thirty-five years before South Sudan.

Importantly, the US provided diplomatic support and humanitarian aid to the SPLM/A throughout its twenty-three years of civil war with the government of Sudan and the US played an effective role in encouraging the government of Sudan to sign the 2005 peace agreement, which ended decades of civil war. Thus, the Government of

---

129     Prunier, G. (2004). Rebel movements and proxy warfare: Uganda, Sudan and the Congo (1986–99). African Affairs, 103(412), 359-383.

South Sudan greatly values its relationship with the US. It knows very well that by being a friend to the US many things benefits are assured. However, what South Sudanese leadership has failed to understand, is that US would be attracted if war is avoided and that basic human rights principles are respected.

Since 2005, the US has spent $14 billion on South Sudan to help fund its development and humanitarian needs.[130] In addition, it advocated for international help for South Sudan's government. The commitment the US has towards South Sudan is a puzzle that academics are still striving to understand. John Young, for example, argues that the Western powers, led by the US, saw Garang as the linchpin of the north–south peace process.[131] Young purports that "indeed, [the] US helped make Garang so and looked to him to resolve the conflicts in western and eastern Sudan, as well serving as a critical element in the 'War on Terror'".[132] As Young argues, "Garang also had close relations with many regional leaders, all of whom hoped he would be instrumental in ending Sudanese support for Islamist terrorism and groups like the LRA".[133]

## The Effect of Diplomacy on the LRA

The effect of diplomacy on the LRA cannot be underestimated. The coming together of Uganda and Sudan against the LRA came as a

130    Amir Idris Why the US must not ignore the struggle for South Sudan's soul. The Hill. 12 May 2018. Available at https://thehill.com/opinion/international/419422-why-the-us-must-not-ignore-the-struggle-for-south-sudans-soul

131    Young, J., 2005. John Garang's legacy to the peace process, the SPLM/A & the south. Review of African Political Economy, 32(106), pp.535-54.

132    Ibid.

133    Ibid.

result of the strong pressure by the US on these two countries. The US, whose main concern was combating international terrorism, felt that defeating the LRA would require the cooperation of both Sudan and Uganda. In 2002, the US labelled the LRA a terrorist organisation and in the subsequent years to come, other international community law enforcement bodies joined in. The International Criminal Court indicted top LRA leader, namely Joseph Kony, on the charge of massive violations of human rights[134]. The ICC took a decision to indict Joseph Kony after it received the case of serious violations, referred by the Government of Uganda in 2003[135].

Cooperation began with intelligence sharing that saw Uganda military intelligence officers being posted into Sudan's Armed Forces barracks located in South Sudan as part of confidence building between the two countries. At the time the intelligence sharing agreement was reached, Sudan had been seen as the main supplier of LRA military equipment.[136] Sudan established contact with the LRA in 1993 through Commander William Nyuon Bany, a breakaway Chief of Staff from the SPLA who surrendered himself to the Sudanese Army garrison in the south, to escape SPLA pursuit after the defection. The link established by the defection of Willaim Nyuon equipped Sudanese forces stationed in the South with insider and helpful information of UPDF -SPLA relations. Sudan took this chance as a justification to

---

134    UN Integrated Regional Information Networks (IRIN), "Uganda: ICC Issues Arrest Warrants for LRA Leaders," October 7, 2005; available at www.irinnews.org/report.asp?ReportID=49420SelectRegion= East_AfricaSelectCountry=UGANDA.

135    Branch, A. (2007). Uganda's civil war and the politics of ICC intervention. Ethics & International Affairs, 21(2), 179-198.

136    Johnson, D.H., 2016. The Root Causes of Sudan's Civil Wars: Old Wars & New Wars. James Currey.

consolidate its military support to the LRA. Now at this point, it was a policy that the LRA would be supported to overthrow Museveni government in Uganda[137]. Sudan's Support of the LRA was intended to encounter Uganda's backing of the SPLA. As Peter Fabricius states: "Sudan's support of the LRA was intended to counter Uganda's backing of Khartoum's foe, the Sudan People's Liberation Movement (SPLM), now South Sudan's ruling party".[138]

Bringing the one-time foes Sudan and Uganda to agree to jointly fight the LRA was a big US diplomatic achievement in the War on Terror. The US was able to compel Sudan to commit itself to an intelligence sharing agreement on the LRA. It was in this way that successful operations against the LRA would be grantee. The success of intelligence sharing meant the killing or capturing of the LRA leadership. Although Joseph Kony, the top LRA commander, has not been killed or captured as projected, the LRA's military activities were sharply reduced to the point of being ineffective, and its activities are now limited to operate in small pockets far from its traditional areas in Northern Uganda, South Sudan and Eastern Equatoria.

Some writers, like Joost van Puijenbroek and Nico Plooijer, have argued that "months into the operation, it became clear the operation was not a success as many people had hoped for"[139]. However, the reduction of LRA field activities, as already alluded, suggests that joint operations against the LRA by the US, Uganda, and Sudan had

---

137     Ibid 113.

138     PETER FABRICIUS. The AU must grasp the nettle before Joseph Kony's Lord's Resistance Army returns to full strength. ISS 06 JUL 2016.

139     Puijenbroek, J.V. and Plooijer, N., 2009. How enlightning is the thunder: A study on the Lord's Resistance Army in the border region of DR Congo, Sudan and Uganda. P. 5.

produced the expected results. It was, therefore, not a failure. However, some challenges related to poor coordination, insufficient funding, lack of training and equipment on intelligence and varying capacities of the partners were noted. It was these shortcomings that contributed to the survival of the LRA till today. Uganda and Sudan relied most on US resources and intelligence capacity, and ignored their own contributions - leading to the total collapse of the LRA pursuit when the US left. For example, their knowledge of the area.

Bringing the LRA's insurgence to a complete end required acknowledging and recognising the role played by each actor - the anti-LRA collaborators namely: the US; Uganda's UPDF; Sudan's SAF and the South Sudan's SPLA. This is the only way for them to achieve their ultimate objective: capturing or killing Joseph Kony and his top commanders, who are still at large. The failure to capture or kill Joseph Kony results from failure to secure accurate information on the intended target. Poor coordination, human error, inadequate intelligence support equipment, shortage of finance, inadequate training, withholding of information, political restrictions and more all contributed to the survival of Joseph Kony, his commanders and the LRA.

Although the US, Uganda and Sudan's operations on the LRA supported by intelligence sharing agreement could be regarded a success because it militarily defeated the LRA by pushing it away totally from Uganda and South Sudan and narrowed its activities to zero operations, nevertheless, chances of the LRA resurfacing are likely, considering the fact that leader Joseph Kony has not been killed or apprehended, and as of today, his whereabouts is unknown. The survival of Joseph Kony is attributed to the fact that the initial intelligence-sharing agreement in 2002 between the Uganda Government and the Sudan

Government failed to produce a situation where intelligence was fully effective in operations against the LRA. It failed to secure accurate information because of poor coordination, human error, and lack of proper intelligence information, as South Sudan's SPLA was never on equal footing with the US and Uganda. Similarly, Sudan, the signatory to intelligence, played politics of double standards. Initially, it was envisaged and perceived that the LRA's criminalisation would put political pressure on the Sudanese government to stop supporting the LRA and to assist in its members' capture.[140] Further, it was conceived that international arrest warrants would isolate the top leadership from the rest of the LRA, making them easy targets once removed.[141] Although not all LRA leaders have been eliminated, international arrest warrants have isolated the LRA internationally and regionally, rendering it ineffective. The indictment of Joseph Kony by the International Criminal Court couple with US pressure compelled Sudan, the main weapons and logistics supplier of the LRA, to cut relations with LRA.[142]

---

140     ICC, "Statement by the Chief Prosecutor on the Uganda Arrest Warrants," The Hague, October 14, 2005, p. 7; available at www.icc-cpi.int/library/organs/otp/speeches/LMO_20051014_English.pdf. See also Tim Allen, War and Justice in Northern Uganda (draft) (London: Crisis States Research Centre, February 2005), pp. 58–59; available at www.crisisstates.com/download/others/AllenICCReport.pdf.

141     Katherine Southwick, "Investigating War in Northern Uganda: Dilemmas for the International Crimi- nal Court," Yale Journal of International Affairs (Summer/Fall 2005), p. 108; available at www.yale.edu/ yjia/articles/Vol_1_Iss_1_Summer2005/SouthwickFinal.pdf.

142     Simonse, S., Verkoren, W. and Junne, G., 2010. NGO Involvement in the Juba peace talks: the role and dilemmas of IKV PAx Christi. The Lord's Resistance Army: Myth and Reality, pp.223-242.

## Conclusion

Attempts to peacefully end the violence and insecurity perpetrated by the LRA across transnational borders of Uganda, South Sudan, CAR and DRC were obviously unsuccessful. All efforts from the governments of Uganda, South Sudan and the US have so far failed. The Uganda Government initiative headed by Betty Bigombe, the former minister for northern Uganda affairs, failed because of lack of political will from both sides of the conflict in Uganda.

Besides lack of political will, there was the ICC indictment of Joseph Kony and other senior members of the LRA. The decision taken by the ICC to proceed with Joseph Kony and other senior LRA commanders' indictment instilled a lot of fear in the LRA leadership. This was probably the greatest factor that surpassed all other factors. In 2003, President Museveni asked the ICC to intervene by investigating LRA abuses, a call that ICC took seriously. Indeed, the ICC prosecutor, Luis Moreno Ocampo, following Museveni's request, declared that there was sufficient ground to investigate the alleged LRA atrocities. Ironically before investigation was completed, the ICC issued warrants for arrests of Joseph Kony and other LRA commanders. Joseph Kony, Okot Odhiambo, Dominic Ongwen and Raska Lukwiya were all charged with war crimes and crimes against humanity.

# Chapter 5

◇◇◇◇◇◇◇◇◇◇◇◇◇◇◇◇◇◇◇◇◇◇◇

# The Military Approach

## Introduction

The failure of the Juba Peace Talks made the military option (which the Uganda government had always favoured) inevitable. This time, the US and South Sudan joined the war against the LRA because both US and South Sudan were affected by the LRA terror. The LRA had been classified as a terrorist organization due to its massive atrocities, thus, attracting the US attention. South Sudan had suffered directly from the LRA operations. The LRA attacked SPLA locations and had also abducted and displaced many South Sudanese, forcing them to cross into Uganda and DRC as refugees or forced them to becoming IDPs in the capital Juba. Jan Egeland, UN Under Secretary General for Humanitarian Affairs and Emergency Relief Coordinator has this to say:

*"Where else in the world have there been 20,000 kidnapped children? Where else in the world have 90 per cent of the population in large districts been displaced? Where else in the world do children make up 80 per cent of the terrorist insurgency movement?...For me the situation is a moral outrage...We hope, on the humanitarian side, that we're now seeing a beginning of an end to this 18year, endless litany of horrors, where the children are the fighters and the victims in northern Uganda...This would take a much bigger international investment – in money, in political engagement, in diplomacy and also more concerted efforts to tell the parties there is no military solution...there is a solution through reconciliation, an end to the killing and the reintegration and demobilization of the child combatants."[143]*

As Jan Egeland clearly and sadly expressed above, the recklessness of the LRA had cost human life, property and much expenditure by the international community in terms of relief provision. Considering the level of damage caused by the LRA, there was no way that the international community would stand aloof and watch people suffering at the hands of gangsters and warlords, whose aims were not for the betterment of human life but the brutish and destruction of humanity. This chapter addresses and examines military strategies that were undertaken to resolve the LRA war.

## The Transnational Nature of the LRA Operations

The transnational nature of the LRA atrocities have been reported by

---

143     United Nations News Centre, "UN relief official spotlights world's largest neglected crisis in northern Uganda," 21 October 2004, http://www.un.org/apps/news/story.asp?NewsID=12297&Cr=uganda&Cr1

many international news agencies, international and non-governmental organizations who witnessed these atrocities happening in Uganda, South Sudan, DRC and CAR. This is not surprising. Some scholars have already written about the transnational character of armed conflicts in Africa. For example, Twagiramungueta, et al, have argued that transnationality is a major feature of armed conflict in Africa, and the most of the so-called 'civil wars' are internationalized[144]. What is alarming with the transnationality of conflict is that it affects a large number of populations which cut across countries.

As the LRA war extended beyond the Uganda border into South Sudan, DRC and CAR, civil population security in these countries was put at risk, as the LRA mostly engaged in attacking unarmed persons. By the middle of 2000, the LRA posed a threat in Uganda and across into South Sudan, DRC and CAR and was already sizably visible. There was already indication that international peace and security was jeopardized. The LRA war attracted international attention and considered the worst in the world in terms of tragedy compared to wars that were happening at the same time - earlier considered to be worst as far as human rights violations were concerned, for example, the war between Lebanon, Israel and Hezbollah and the fighting and massacres which took place in the Sudan's Darfur region. These conflicts seized the world's attention [145], but the LRA atrocities drew more attention.

In this regards, Abou Moussa, Special Representative and Head of the United Nations Regional Office for the Central Africa region

---

144    Twagiramungueta, et al, Re-describing transnational conﬂict in Africa, p. 378.

145    Seybolt, T.B., 2007. Humanitarian military intervention: the conditions for success and failure. SIPRI Publication. P.IIX.

(UNOCA), pointed out that a sustained international focus was needed to eradicate the multitude of threats that had blighted Central Africa for years — from the security crisis in the CAR, to piracy in the Gulf of Guinea, to the unrelenting terrorist threat posed by the LRA.[146] He went on to say that UNOCA must continue to lead in guidance in states deliberations on sub-regional security challenges. He further argued that while joint diplomatic efforts by the African Union Special Envoy for the LRA Issues had ensured cooperation among the countries affected by that group's activities nevertheless the momentum should be maintained.[147]

## LRA as an International Terror Organization

The US government categorizes LRA as a terror organization due to its use of systematic violence to achieve political objectives. The LRA war has altered lives of millions of Ugandans, South Sudanese, Congolese and Central Africa citizens, because it has forced many of these citizens to leave their homes. As Marieke notes:

> ...the LRA has fought this war with ruthless attacks and abductions, and the Government of Uganda has responded with structural violence on a grand scale against the people of northern Uganda... Warfare tactics on the government side consisted of forcing the entire population in these areas into so-called protected villages, which are in

---

146    See Security Council SC/11182. Briefing Security Council, Senior Envoy in Central Africa Calls for Sustained International Focus on Eliminating Lord's Resistance Army, Other Threats 20 November 2013 available at https://www.un.org/press/en/2013/sc11182.doc.html

147    Ibid.

*reality displacement camps with inhumane conditions...*[148]

The LRA use of brutality, distinguishes it from its predecessor the Holy Spirit Movement (HSM) which Alice Lakwena led. This brutality was seen in the LRA reaction to the Acholi people when they formed their local defence units. As the LRA intensified its brutality on the local population, the Acholi population organized their Local Defence Unit (LDU) known as the Arrow Group. The LRA in return, responded by wide-spread killing, mutilation and abduction of youth and forcefully compelling them to join its rank and files. In this regard it is argued that comparably, the HSM led by a self-claimed prophetess Alice Lakewena, was more accountable than the LRA led by Joseph Kony. Lakewena denounced conscription and tolerated opponents[149].

While the LRA is widely known for its human rights abuses, which are openly a violation of International Humanitarian Law, also known as Laws of Armed Conflict (IHL/LOAC), the Uganda government's forces were equally accused of similar abuses. Among major criticism levelled against the Uganda military has been the forced displacement by relocating Acholi people from their homes into protection camps, where it was alleged that killing and raping of civilians was conducted by the military [150]. This point is further elaborated by Christian Aid which notes: "other allegations which are war crimes that have been labelled against the government forces include extra-judicial killings, torture, and forcible displacement of over one million civilians[151]".

---

148    Ibid. p. 10.

149    Museveni, p.260.

150    Ibid 261.

151    See Christian Aid, 'Background to ther Crisis', available at http://www.christianaid.org.uk/uganda/background.htm.Also Apuuli. P.62.

This was not expected from Uganda because Uganda is a signatory to the ICC and is under obligation to stand against war crimes by punishing those involved. Furthermore, as a signatory to the Convention against Torture (1984), Uganda is obliged not to condone the crime of torture under any circumstances, at any time anywhere within the territory of Uganda or abroad. Prevention of torture is a rule of customary international law, which prohibits and binds every state to outlaw and punish torture.[152]

The LRA has been known for its policy of rape, abduction of children and women. Protocol II - the Additional to the Geneva Conventions of 1977, prohibits all recruitment into forces of children under the age of 15[153]. The prohibition is binding on regular and irregular forces and if violated, there are consequences. For example, on consideration of violation of IHL, in 2005 the ICC issued arrest warrants on Joseph Kony, Vincent Otti, and three other LRA senior commanders for their roles in human rights violations[154]. The case of the Aboke Girls, highlighted below, is one of many well documented crimes against humanity that Joseph Kony's LRA perpetuated.

## The Realist Justification for the Fight Against the LRA

Arguably, the US involvement in the fight against the LRA was driven and guided by the Realist theory of international politics. The Realists view the world that we live in as a world of opposing interests and of conflict among them. The US view on the LRA, is that it is a terror

---

152    Robertson, Geoffrey. 2002. Crimes Against Humanity: The Struggle for Global Justice. Harmondsworth. Penguin Books. P.103.

153    Article 4(3) of Protocol Additional to the Geneva Conventions of 12 August 1949.

154    Mareike Schomersus (2007). The Lord's Resistance Army in Sudan: A History and Overview. Small Arms Survey.

organization which threatens the US security. Therefore, the US must fight the LRA by all means to protect its citizens and their properties. The LRA connection with Sudan and equally its dealing with the civil population qualifies it to be terror organization.

Until 2021, the US had listed Sudan together with Iraq, Libya, and South Yemen as among those states that are sponsors of terror[155]. Sudan was added to the list 21 years ago, when the US concluded that Sudan harboured members of the Abu Nidal Organization, Hezbollah and Islamic Jihad. Bin Laden, the mastermind of the 9/11 US World Trade Centre massacre, had spent half a decade there before he was expelled from Sudan in 1996. US analysis was that Bin Laden's time spent in Sudan were reasonable and crucial for the development of his terrorist network[156]. Further, the LRA engagement in the perpetuation of enormous crimes against the civilians in northern Uganda and beyond the Uganda borders produced significant advocacy from the US based groups who sought US involvement to bring an end to the conflict in Uganda. Consequently, in 2009 the two houses of the US Congress discussed the LRA crisis, and the disarmament of the LRA was signed into law by President Obama. The LRA abuses included abductions, killing, torture, rape, forced displacement, attacks on schools, hospitals, religious sites, IDPs camps. All these acts in addition to its connection with Sudan made the US government to designate LRA a terrorist organization.[157]

---

155    States Sponsors of Terrorism is a designation applies by the United States Department of State to countries which the Department alleges to have repeatedly provided support for acts of international terrorism.

156    Astill, James. Osama: The Sudan Years. The Guardian, Tuesday 16 Oct 2001

157    Mareike, p. 15.

## The Liberals Justification for the Fight Against the LRA War

The United States being a Liberal country that is founded on the political and social philosophy that promotes individual rights, civil liberties democracy, and freedom was the first stand to denounce LRA human rights violations in northern Uganda, and was quick to join the the fight against LRA. It is a common knowledge that liberals do not like wars, they believe in peaceful solution of the conflict. Liberals opposition to wars has a reason. Wars are destructive and bring about human suffering and deny them safety. Unfortunately, the LRA approach to conflict resolution is contrasts greatly to that of liberal theory. Their solution to conflict is to resort to violence. The LRA, from the time of its inception, has committed serious atrocities on those who do no sympathise with them. They were cast as unwanted among the Acholi community, who Kony took for granted as supporters. This is against liberal principles of freedom of opinion.

The US intervention in the Ugandan conflict can be explained within the principle of liberal theory which advocates for human liberties, personal dignity, free expression, religious tolerance, and universal human rights among other good things. Equally, Neoliberalism, like its father liberalism, advocates for civil rights which includes the protection and privileges of personal liberty for all citizens.

When these rights are challenged, international peace and security is also challenged. Following the 9/11 disaster, the US defined Bin Laden and its organization Al-Qaida as a threat to international peace and security because of its terrorist nature. A terrorist organization possesses certain attributes. As Sullivan describes:

*Contemporary terrorism is a complex phenomenon involving a range of non-state actors linked in networked organizations. These organizations, exemplified by the global jihadi movement known as al-Qaeda, are complex non-state actors operating as transnational networks within a galaxy of like-minded networks. These entities pose security threats to nation states and the collective global security.*[158]

Indeed, these organizations pose a threat world-wide because of their network coverage. The terror attack on the World Trade twins' towers was blamed on the Al-Qaida network, which caused the US government to authorize Afghanistan invasion. As Okoth explains:

*...after attack of the World Trade Centre and Washington, the US administration undertook an invasion of Afghanistan shortly after the September 11, 2001*[159].

The US government decision to invade Afghanistan was intended to uproot Al Qaida, by uprooting the Taliban government which served as Al Qaida breeding ground. As Okoth further notes:

*... the main motive of the US was to remove from power, the Taliban led government that has been accused by the US as providing safe*

158    John P. Sullivan, Terrorism Early Warning and Co-Production of Counterterrorism Intelligence, Canadian Association for Security and Intelligence Studies CASIS 20th Anniversary International Conference Montreal, Quebec, Canada 21 October 2005.

159    Pontian Godfrey Okoth. Ed. (2008). Peace and Conflict Studies in A Global Context. Masinde Muliro University of Science and Technology Press.p.17.

*haven for Osama bin Laden, the leader of al-Qaida[160].*

Indeed, in its annual routine publication of countries sponsoring terrorism Washington listed Afghanistan alongside with Sudan, Cuba, Iran, South Korea, and Iraq as the countries encouraging terrorists' attitudes[161]. 'Terrorism' as defined by Graham Evans and Jeffrey Newnham in the Dictionary of International Relations, is the use or threatened use of violence on a systematic basis to achieve political objectives.[162]

## The Uganda Military

The National Resistance Army, later renamed the Uganda People Defence Forces (NRA/UPDF), pursued a military approach from the very beginning of the war in northern Uganda. Museveni had just ascended into power and was still energetic to fight more wars and show his descendants that the NRM had different approach to solving Uganda problems. Under his guidance, the war on LRA was launched to defeat Kony by capturing him alive or dead. This move was supported by series of military operations, which are discussed in detail below.

---

160     Ibid.

161     Frederick H. Careau (2004). State Terrorism and the United States. Zed Books. P. 15.

162     This definition is given by Graham Evans and Jeffrey Newnham. For details see Graham Evans and Jeffrey Newnham (1998). Dictionary of International Relations. Penguin Books. P.530.

## Operation North

The first major military offensives launched by the Uganda armed forces (the NRA by then) to stop Joseph Kony from attacking civilian locations, was an operation codenamed Operation North. It was launched in March 1991. As part of preparation to enhance success against Kony, all of Northern Uganda was locked down. This lockdown exercise affected the humanitarian organisations that were providing services among the IDPs in northern Uganda because they were forced to leave the area[163]. As part of operation security, media coverage was halted. As the offensive progressed, government forces taking part in the operations were alleged of committing serious human rights violations against civilians, violations which included torture of suspected prisoners among the Acholi people, most living in the regional city of Gulu. Also, it was alleged that the Uganda military engaged in extra-judicial killings and detention of prominent Acholi leaders.[164] It was further alleged that a portion of these violations were done by the local militia - the home guards known as Local Defence Units (LDUs) mainly recruited by the military from among the Acholi, whose main task was to identify the LRA members and their collaborators and report to the forces pursuing the LRA.

## The Local Defence Units (LDUs)

The genesis of this local militia goes back to the then Uganda minister in charge of the north, Betty Bigombe, who is Acholi herself. She initiated a Local Defense Unit (LDU) composed of locally recruited Acholi young men, who were mostly armed with bows and arrows,

---

163    Prudence Acirokop., p. 8.

164    Ibid.

to bridge the security gap where the Uganda military and the police failed. LDUs were charged with duties of protecting villages gazetted by the Uganda military. Local defence units are common during wars and are used as counterinsurgency units. During the Vietnam War, the CIA created Village Defense Program in South Vietnam as counterinsurgency program[165]. In Afghanistan, local militia have played a significant role in the aftermath of Taliban incursion[166].

The creation of the LDUs in the Acholi land that Joseph Kony had assumed to be his strong hold led him and his organization to believe that it had lost the population's support to the advantage of Museveni. He blamed the Acholi civil population for shifting allegiance and held them responsible for their cooperation. He therefore escalated reprisals and intensified killings without distinction, aiming to terrify people so that they would abandon dealing with government, forgetting that the more brutality is applied, the more ground is lost. To make sure that Acholi people were scarred, Joseph Kony ordered his combatants to engage in dehumanizing its own people. These actions included cutting off his victims' lips and killing others by hacking them, by using machetes, the weapon applied or used to perpetuate the Rwanda genocide in 1990. Also, the LRA disrupted development in the region and denied access to health care; and medical treatment provided by the hospitals, resulting in closure of health facilities due to lack of medicines and the departure of health personnel, who left to where their security was guaranteed. Similarly, education was disrupted too. Like their colleagues in the medical field; teachers left northern Uganda

---

165    Strandquist, J., 2015. Local defence forces and counterinsurgency in Afghanistan: learning from the CIA's Village Defense Program in South Vietnam. Small Wars & Insurgencies, 26(1), pp.90-113.

166    Lefèvre, M., 2010. Local Defence in Afghanistan. Afghan Analysts Network.

because their life was at stake. This led to schools being closed down. On the side of girls, they were forced into early marriages and pregnancy imposed by appalling economic conditions. These conditions prevailed until Operation North, launched by the Uganda military, came into existence. Operation North succeeded in compelling Joseph Kony and the LRA to retreat and cross into southern Sudan[167], thus relieving northern Uganda from the burden of the LRA but shifting it to southern Sudan.

Following this military achievement, the Uganda government requested the Government of Sudan to allow it pursue Joseph Kony to their territory. Thus, on 27 April 2002, Sudan and Uganda agreed to establish a joint ministerial committee for the resumption of full diplomatic relations, which had been terminated by Uganda in 1995 in protest against Sudan's connection with the LRA. Due to this diplomatic move, relations between these two countries briefly began to ease, and the immediate result of this, was the resumption of tripartite negotiations involving the governments of Sudan, Uganda and the LRA. Accordingly, the LRA temporarily halted its attacks on northern Uganda to allow the effort develop. The agreement permitted Uganda forces to hunt the LRA inside Sudan for couple of months and to end in May, 2005. As per this arrangement the Uganda forces launched a hasty massive military offensive - Operation Iron Fist, against the LRA bases in South Sudan.

## Operation Iron Fist

The missionary news service of the US reported on May 9 2002, that 10,000 Ugandan soldiers were in Eastern Equatoria, southern Sudan,

---

167     Southern Sudan was still part of Sudan by then.

approximately about 60 kilometres from the Sudan and Uganda border, heavily equipped with tanks and artillery[168]. This report was the first broadcast of the government's move towards the LRA. Correctly, these troop movements were meant to hunt down Joseph Kony and his LRA supporters. Kony relocated his bases from Uganda into South Sudan to escape mounting Uganda military pressure and also to provide protection shield to the Sudanese army against SPLA attacks, the condition that the LRA must fulfil in exchange of military logistics. The LRA launched its operations into Uganda from the Southern Sudan border, prompting Uganda's government to send its operations into Southern Sudan[169].

There was another unique characteristic Operation of Iron Fist. Operation Iron Fist was part of the US declared War on Terror. In December 2001, in response to the request of the Uganda Government, the US government placed the LRA in their list of terrorists' organizations[170]. US placement of Joseph Kony movement among terrorist organizations gave the Uganda government a boost. Soon after this declaration, the Uganda People's Defence Forces (UPDF) launched the Operation Iron Fist (OIF). The OIF mission, as I already briefly mentioned above, was to chase the LRA from their hiding place in southern Sudan, destroy the logistical bases, and kill or capture Joseph Kony. However, the success of this operation depended on the explicit

---

168 News from Africa, 'Uganda and Sudan join hands to fight LRA', November 11, 2009.

169 South Sudan became independence in 2011 following a referendum reached in the Naivasha, Kenya Comprehensive Peace Agreement (CPA) between the SPLM/A and Government of Sudan

170 Rodriguez, C., 2004. The Northern Uganda War: the "small conflict" that became the world's worst humanitarian crisis.

support of the government of Sudan. The government had to support to change its face in the eye of the US, which, had earlier indicated its retaliation if the government of Sudan did not cooperate with the government of Uganda against the launched war on the LRA. Following 9/11, the US had accused and listed Sudan in the list of those countries cooperating with terrorists. Sudan promised its cooperation. Unfortunately, OIF became catastrophic for the people of northern Uganda because Kony responded mercilessly by sending into northern Uganda his forces which committed unbelievable range of further atrocities than in previous years. Abductions, particularly of children and displacement became common and more unfortunately, the war spread into Lango and Teso, regions which had remained calm before the launch of Operation Iron Fist. In order to contain the situation, the government responded by creating some local militias.

As Operation Iron Fist continued, voices of criticism emerged. They argued that the Operation Iron Fist Iron was launched by Museveni to exploit the souring relations between Sudan and US. In their view, the operation was not about eliminating Joseph Kony as Museveni made public believe, but rather about Museveni being anxious to secure relations with the US, as relations between Washington and Khartoum began to deteriorate. They based their argument on the speech Museveni delivered on 6 May 2011. During his official visit to US, Museveni gave a speech at a reception organized by his American and Ugandan supporters in Washington, D.C. It was during this reception, that Museveni was alleged to have expressed his cooperation with the US intelligence service in order to fight terrorism.[171]

Museveni and the US had labelled Joseph Kony and his movement,

---

171     http://africa.peacelink.org/newsfromafrica/articles/art_908.html, May, 2002.

the LRA, a terror organization especially following devastating reaction to Operation Iron Fist. LRA operations mobilized the international community to turn against it with almost majority of them promising to stand by Museveni in his declared war against the LRA. Without brutal activities of the LRA, the conflict would have not gained remarkable and unprecedented international concern. The international community became part of the war on the LRA because they were frustrated by the LRA abuses on civil populations. Such abuses were ascertained during field visit of high profile. For instance, in November 2003, the then United Nations Undersecretary General for Humanitarian Affairs and Emergency Relief Coordinator, Jan Egeland, undertook a field visit to Uganda. Upon returning, Egeland testified that LRA was a real threat to international peace and security. His testimony brought international attention to the conflict[172]. Egeland's visit to the LRA areas of atrocities persuaded international community to pledge to Uganda increased humanitarian funding from 19.5 million US dollars to 56 million US dollars by 2007. In addition, total official development assistance and the aid as well, increased from 817 million US dollars in 2000 to 1.2 billion US dollars in 2005[173]. The doubling of the development and humanitarian assistance demonstrates the world's commitment to see to it the LRA was defeated.

The US and the international community - commitment as shown by the financial commitment, did not build confidence among the Acholi people who Kony and LRA war had affected greatly. For

---

172    See also 'Uganda Conflict Worse than Iraq' BBC News 10 November 2003 http://news.bbc.co.uk/2/hi/africa/3256929.stm (accessed 16 January 2009.

173    See Uganda Data Profile (2007); UNOCHA Financial Track Service: Uganda (2000), (2002) & (2007).

example, the Acholi leaders were not hopeful that Operation Iron Fist would improve their situation or guarantee security for northern Uganda. The reason for this concern was that the biggest victim of the Operation Iron Fist was abducted children, as in the previous operations. Causing further concern, were reports that young boys and men were being forcibly recruited by the Ugandan military into its forces to fight the LRA.[174]

The concerns of the Acholi people with regards to suffering of their people were reasons in the first place why international community and the US became interested in Uganda internal affairs. They indeed wanted to stop Kony from deepening the worsening situation in northern Uganda, thus they took their own initiatives to bring about an end to the child abuse that LRA was perpetuating. The LRA abducted children and recruited them into their service. In the battlefield, they became vulnerable to the Uganda military. They were the first to be killed in Ugandan military operations against the LRA.

It was alleged that the Ugandan military was not taking prisoners, nor attempting to secure the safety of abducted children, which was one of the main reasons Uganda military justified entering into South Sudan. UNICEF Executive Director, Carol Bellamy told the press that UNICEF had not seen any evidence to date that the children were being rescued. This is what he had to say:

> We need to find out where these children were and then do everything possible to ensure their protection and ultimately, reunification with their families... While it may be necessary to use some degree of force in

---

174     Linda Fromemer, "Uganda and Sudan join hands to fight LRA ', May 2002 in http://www.newsafrica.org/article/art

*preventing an armed attack, we cannot forget the LRA was made up at present of at least 70 percent abductees, mostly children and that a direct attack against their bases would end in the destruction of many innocent lives[175]*

Archbishop Odama of Gulu shared similar views. In an interview with Uganda daily newspaper The Monitor, he remarked that:

*In fact, recent figures of rebel casualties give us no reason for joy, since we have learnt that most of those who died were children who could have been reintegrated, as many of their companions have been. He continued to say: "it is my conviction, as well as that of the vast majority of the people of Acholi, that our 16-year-old conflict will be ended by peaceful dialogue. Nobody, whatever his position or side, must refuse this avenue[176]*

It is obvious dialogue and negotiations were the best option to end the LRA war. However, as much as voices were calling for peaceful resolution of the conflict over those calling for a military solution, the reality or the focus should be on Joseph Kony. His determination for peace would help the whole process succeed, thus, military options would be avoided. In early April 2002 Acholi elders held a conference where they openly denounced intensified conflict with the LRA. They argued that intensified conflict with LRA would lead directly to the death of Acholi children or that they would become lost in the forests

---

175    Carol Bellamy , Press Conference released by Unicef, Kampala, 25 May 2002.

176    Archbishop John Baptism Odama, statement at Regional Conference on Cross Border Peace building, Gulu, May 2009.

of Southern Sudan and die. The elders called upon the Uganda government to follow up its amnesty offer more vigorously as a way to pursuing an end to the war with the least loss of life.

Of more interest to the Acholi leaders, was the situation of their IDPs in the Uganda forces protected zones. The Acholi community at large felt that the protected villages must be immediately closed down and that the government must an enable the people in those protected villages to return to their homes and protect them from LRA attacks from there. This view is affirmed by the statement of an Acholi Roman Catholic Priest Odongo who told AFRICA NEWS that:

> *The government of Uganda should dismantle the camps so that the people can go home. Once the camps are dismantled and the government guarantees the security of the people, there is need to financially empower the people so that they can begin to fight poverty and diseases… There is urgent need to rebuild the infrastructure of northern Uganda: roads, schools, health centers, etc…Most parents in northern Uganda are unable to pay for the tuition fees of their children…there is need to establish special funds to cater for the education of children in northern Uganda… This will cover the orphans and children whose parents have been affected by the war.[177]*

In line with such comments, whether or not the government pursued its military option on the LRA, people must be protected from the LRA. As part of its protective measures the Ugandan government adopted the policy of protected villages following several attacks on local population villages by the LRA, to provide security to the victims

---

177     Msgr Mathew Odong, in http:// www.newsafrica.org/articles/art

through guard of the UPDF. However, UPDF protection was soon to be criticized as allegations of civilian harassments by the UPDF began to surface. Although a degree of security prevailed, it did not stop LRA operations. The continued LRA attacks led to the second round of operations, this time with different name - Operation Lighting Thunder.

## Operation Lighting Thunder

As part of the follow up mission to completely deal with LRA atrocities, the Uganda government launched a third major campaign of operations after the failure of the earlier ones. This time however, the Uganda government sought external help from neighbouring countries affected by the LRA. Thus, in December 2008, the armed forces of Uganda (UPDF), the DRC *Forces Armées de la République Démocratique du Congo* (FARDC) and South Sudan (SPLA) with military and logistical support from the United Nations peacekeeping mission in Congo (MONUSCO) and the US Africa Command (AFRICOM), launched a joint military offensive on the LRA. Launched 6 years after the failure of the Iron Fist Operations, Operation Lightning Thunder aimed to root out the LRA rebel group that had become a regional problem rather than being just a Ugandan problem.

Earlier, Uganda had launched a number of military operations against the LRA in an attempt to defeat it militarily. The outcome of such military operations was pushing away LRA forces, now compelled to cross into South Sudan, DRC and CAR. The effect of Operation Iron Fist on the LRA was immense but did not achieve a full victory. The LRA continued to attack the civil population across the borders of South Sudan, DRC and CAR. As Sylvester B. Maphosa writes:

"In 2002, the GoU launched Operation Iron Fist undertaken by

Uganda People's Defence Force (UPDF), i.e., the national army.... The operation ignited intense and violent attacks in retaliation by the LRA in northern Uganda several thousand families were dislocated, and the operation in all intents failed to end the hostilities."[178]

## Foundation of Success for Operation Lighting Thunder

What is known comes from the past. There are many lessons to be learned by the Operation Lighting Thunder, as compared to Operation Iron Fist and its predecessor, Operation North. The paramount and most important thing to be learned from these previous operations namely Operation Iron Fist and Operation North, is the reaction of the LRA when confronted by Uganda government forces. In other words, the Uganda and Allied forces battling the LRA must have known that the reaction of the LRA towards their moves would be the same as in previous operations. It must be predicted that the LRA response to Operation Lighting Thunder, would be to brutally attack civilians as in the past. Thus, the immediate actions that Uganda, South Sudan, DRC and CAR should have taken was to consider how to protect the civilians against the LRA attacks. Indeed, when the UPDF, supported by the SPLA, launched Operation Lightning Thunder on 14 December 2008 by bombing the LRA'S camps inside Southern Sudan to pressurize Joseph Kony to surrender and sign the agreement, the LRA immediately resorted by attacking civilians. It retaliated with brutal massacres. The massacres became the point of discussion and the military efforts were quickly buried. The affected people and their

---

178     Sylvester B. Maphosa. The Lord's Resistance Army: A Review of African Union Regional Efforts to Eliminate the Resistance in Central Africa in Festus B Aboagye Ed. The Comprehensive Review of the African Conflicts and Regional Interventions, 2016. P.193.

sympathisers were the quick to blame government of Uganda and its forces -the UPDF. Critics argued that LRA attacks were increased because of the presence of army in their region. They argued that it was the presence of the army that left them vulnerable.

Resulting from LRA increased attacks on civil population, a considerable number of people were forcefully displaced to new places where even limited access to humanitarian aid was difficult. Thus, from the eye of the normal innocent Acholi Ugandans and the South Sudanese affected by the LRA revenge attacks Operation Lightning Thunder was a burden to them rather than being a salvation to them. However, if we are to give the devils its due, Operation Lighting Thunder may be to a large extent be considered a success and a failure at the same time. It is a success because the LRA has no presence in Uganda at the time of publication and its operations have been silenced in South Sudan for the last couple of years. The LRA is now hiding in the bush, in DRC and CAR.

The LRA, from their hide out places in DRC, CAR and South Sudan, continued to conduct highly coordinated ruthless attacks against civilians. By mid-December 2008, the LRA had brutally murdered more than 1,000 people in north-eastern Congo and southern Sudan and abducted nearly 250 children[179]. In at least one case, in north-eastern Congo's Orientale province an entire village was privileged and burned to the ground and more than 180,000 Congolese were forced from their homes, while in Southern Sudan, a further 60,000 were displaced[180]. This happened because of the poor gathering and lack of timely intelligence sharing among the combined

---

179    AFP, February 10, 2009.

180    http://www.enoughproject.org/publications.

forces, and insufficient logistical support. These factors contributed to the inability of Operation Lighting Thunder soldiers to fulfil their protection task efficiently. The result, was IDPs were highly vulnerable in terms of physical and food security, as humanitarian access in these remote areas was limited. Without proper and timely and sharing of intelligence by the forces pursuing the LRA units, a golden opportunity to defeat the LRA was lost. If the LRA is not defeated and remains at large as it is now, and without the spirit to genuinely seek political settlement, it is impossible to suggest that more death and destruction, especially when the LRA uses the current space and time to reorganize and to rebuild its military strength, will not repeat itself. The longer intelligence gathering and sharing is delayed, the more time it gives the LRA to regroup, opening another wave of further havoc. The UPDF and SPLA were able to destroy some of Kony's main camps; and rescued a sizable number of abductees; and killed some rank-and-file among Kony fighters. However, these gains were not sufficient to bring an end to the LRA operations. The LRA leadership is still at large.

Finally, the presence of Operation Lighting Thunder soldiers in southern Sudan and north-eastern Congo temporarily improved security in some areas. Every person interviewed in and around the Doruma and Dungu areas of Haut-Uele district in Orientale Province said that the Operation Lighting Thunder soldiers should not withdraw until they have successfully captured or killed Joseph Kony and the senior leadership of LRA.[9] Indeed, Aljazeera English News reported killing on 29 civilians by the Uganda rebels presumably LRA[181].

This inability of the forces to defeat and get rid of Joseph Kony

---

181    Aljazeera English news 1 April 2021

makes people in Uganda, South Sudan, DRC and CAR and even Sudan Western Darfur region nervous about the uncertainty of their security now and in future. Since the launch of Operation Lightning Thunder, the LRA intensified attacks on civilians, killing and abducting the most vulnerable such as children, sick and elderly. Because UPDF and SPLA forces have failed to protect civilians adequately, the local population in those countries affected by the LRA rebels are now left with no choice but to organize themselves into local militias in form of self defence units. The LDUs come with their own problems. For example, the United Nations Mission in South Sudan (UNMISS) released a report stating that community-based militias were responsible for 78 percent of killings and injuries caused to civilians, as well as abductions and conflict related sexual violence during attacks in pockets of South Sudan[182]. Previously, these groups of armed civilians, for example the "Arrow Boys" in western Equatoria, formed themselves to fill the security gap in the area and protect themselves from the LRA attacks. In some instances, for example as in the town of Bangadi, Western Equatoria State, LDU effectively pushed back the LRA. While it is a good thing for people to protect themselves, these local militias - as in the case of Arrow Boys, were converted into rebels which perpetuated insecurity in area.

Numerous dangers are associated with these informal forces. None of them have received any training. They do not operate according to any codes of conduct, and they do not have to answer to higher authority. While they have played a role in protecting civilians and providing intelligence on the LRA to the SPLA, and UPDF, there is a

---

182    UNMISS, Community based militias responsible for 78% of victims of violence in South Sudan, 31 March 2021.

significant risk that they will organize and fashion themselves into the new predatory groups in future, as has happened repeatedly in Congo, unless the SPLA and UPDF make efforts to bring them under their control.

## Funding Operation Lighting Thunder

Although Uganda UPDF and South Sudan SPLA were determined to collaborate fully as equal partners to address the LRA shared security threat, in reality it was the US government funding that was the driving force behind the execution of Operation Lighting Thunder. However, the US funding was focused and concentrated on the UPDF. The SPLA, despite immense contribution in the operation, was not budgeted for. This support provided to the UPDF included interception equipment which an enabled UPDF to intercept LRA communications such as satellite phones. it also allowed UPDF to triangulate LRA positions. Also, the US military advisors stationed in Uganda as part of the support to UPDF provided the UPDF with satellite imagery and maps.

The US one sided funding imbalanced the operations. The SPLA, with its area and terrain knowledge, was hindered by the lack of supporting intelligence gadgets, thus the survival of the LRA. In an interview with SPLA Chief of Staff, he acknowledged that Southern Sudan border is simply too large and too unpopulated, and that his forces were not equipped to actually be able to monitor and seal the border[183]. He added that the LRA has been able to exploit this operational weakness by moving back and forth across the Congo-Sudan border,

---

183     Interview with Gen. James Hoth, Juba, 6 August 2010

evading capture and targeting civilians[184].

This prompted US Africa Command (AFRICOM), to review the operational plans, and provide advice on its execution. Unfortunately, the civil war that erupted in South Sudan in December 2013 discouraged the US. The end result was the US withdrawal of funding and that was the end of Operation Lighting Thunder. Prior to the US funding withdrawal there were also been allegations that AFRICOM did not make available the planning capacity to offer substantive support and advice during the planning and execution phases of Operation Lighting Thunder. It was further alleged that when the operation encountered significant difficulties, US officials disassociated themselves from the operation.

Additionally, territorial restrictions imposed by the neighbouring Congolese government seriously limited the UPDF and the SPLA ability to track the LRA in DRC. For example the Congolese army forced two Ugandan companies to leave Faradje when LRA activities were concentrated in Faradje. This restriction was probably attributed to lack of trust and confidence between the Uganda UPDF involved in Lightning Thunder and the Congolese military. Ugandan, Congolese, and UN officials all expressed that critical intelligence was not shared amongst the forces. This reluctance to fully disclose information limited the extent to which the armies worked together in a variety of areas.

Given the US role in the LRA fight, the Joe Biden administration now has a responsibility to help finish the job of finishing the LRA war. US involvement will also be critical to ensuring that the Uganda government and its army do not stray from the mission of defeating the LRA. Pursuits must continue. Further, international reassurances

---

184      ibid

and engagement are absolutely critical to keeping mutual the regional countries consisting of Uganda, South Sudan, DRC, CAR and Sudan on track at ending the LRA threat. With the support of US planning, intelligence, and logistical capabilities, there is no doubt that the LRA will be defeated, thus saving civilians from more sufferings.

## The US Special Forces

In his forward to Lee Kuan Yew, *From Third World To First* Henry A. Kissinger writes that "At the same time, technology has made it possible for nearly every country to participate in events in every part of the world as they occur"[185]. Indeed, in 2010 having been influenced by international media which publicized the LRA's evil treatment of civil population in northern Uganda, President Barak Obama reacted to the situation by pushing and signing into law, the LRA Disarmament and Northern Uganda Recovery Act of 2009. Essentially, the policy endeavor to support the stabilization and lasting peace in northern Uganda and areas affected by the LRA through development of a regional strategy to support efforts to successfully protect civilians and eliminate the threat posed by the LRA. Accordingly, in October 2011, the US government deployed a team of around 100 military intelligence experts to Uganda, South Sudan, CAR and the DRC to assist them in their endeavor to defeat the LRA insurgency. The policy was also intended to apply other methods such as encouraging defections from the LRA, and protecting civilians through intelligence sharing. This is not a first time that foreign governments and the US in particular, intervened to address human rights violations in other countries.

---

185    Dr Henry A. Kissinger in Lee, K.Y., 2000. From third world to first: Singapore and the Asian economic boom. New York: Harper Business. P. 8

The US intervened to stop violations in Darfur by the Government of Sudan militia. As Seybolt remarks "many governments and the United Nations (UN) have echoed the concern, with the United States going so far as to officially accuse the Sudanese Government of genocide".[186]

## The SPLA

A good thing done yesterday pays tomorrow. Salva Kiir support to Museveni in his fight against the LRA justified Museveni's intervention in the conflict that erupted in South Sudan between Salva Kiir and Riek Machar in December 2013. Museveni sent a contingent of well-equipped UPDF to halt Machar's rebel advancement towards the capital in Juba in December 2013 on the ground that it was for humanitarian protection of Ugandan citizens in South Sudan.

It is undoubtedly that for many years both SPLA and the LRA fought proxy wars between Uganda and Sudan in exchange of war logistics. As already stated above, in 1987, the LRA founded on spirituality by Joseph Kony, became a de facto force in northern Uganda aspiring to establish theocratic government in Uganda as soon as it overthrows President Museveni's National Resistance Movement (NRM) government. The northern Uganda rebellion stemmed from enduring colonial legacies of political and ethnic divisions between the northern and southern parts of Uganda[187]. However, this historical explanation on the cause of the LRA war is negated by other arguments

---

186     Seybolt, T.B., 2007. Humanitarian military intervention: the conditions for success and failure. SIPRI Publication.

187     Kihika, KS. Evaluating the Deterrent Effects of the International Criminal Court in Uganda. In Schense, J. and Carter, L. (2016). Two Steps Forward, One Step Back, Deterrent Effects of the International Criminal Tribunals. International Nuremberg Principles Academy. P. 204.

which suggest that the LRA to have emerged as a proxy war that the Sudanese government of Field Marshall Omar Hassan Hamed Al Bashir agitated in retaliation for the alleged Uganda government support to the SPLA[188]. Indeed, such conspiracy theories are not uncommon in the African Continent. As Salehyan argues:

> ... *countries in the Horn of Africa tend to seek solutions to their problems by providing sanctuary to rebel groups from neighbouring states, and they intervene in the affairs of their neighbors to counteract perceived threats to domestic interest.*[189]

Sudan and Uganda, from the time of their independence from Britain have experienced internal rebellions. In Sudan, the southern rebellion- the ANYANYA Movement a political organization formed by the Southern Sudanese who rebelled against the central government in Khartoum in 1955 from among the Sudan military forces based in the south, became a source of conflict. Sudan government accused Uganda of hosting political refugees, an accusation which Uganda denied. There were other disputes between them over demarcation and claims of borders along Kejkeji, Nimule and Magwei corridors. Following the eruption of civil war in Uganda following over throw of General Amin Dada in 1978 and subsequent flee of Amin supporters to become refugees in Sudan, the same accusation of hosting political

188    Frank Van Acker. 2004. Uganda and the Lord's Resistance Army: the New Order no one ordered African Affairs, Vol. 103, No. 412 (Jul., 2004), pp. 335-357.

189    Salehyan, I. 2009. Rebels Without Borders: Transnational Insurgencies in World Politics. New York: Cornell University. Quoted in Samson Wassara. Why Conflict in South Sudan and Somalia is Beyond Prevention and Management. Africa Insight Vol. 49(3), December 2019.p.104.

refugees emerged this time with Sudan being the accused. In 1983, civil war in Southern Sudan erupted for the second time after one decade of guns silence brought by the Addis Ababa Agreement, 1972.

## Conclusion

The Ugandan government has launched several military operations against LRA. Since 1986, the government has launched series of military offensives against the LRA. The first was Operation North, in 1991. Operation North succeeded in weakening the LRA and, by 1992 and 1993, the intensity of the conflict was greatly reduced. Following normalization of relations between Bashir and Museveni governments, an agreement to allow the UPDF to pursue the LRA across the Sudanese border was established. The first large-scale military operation under this truce, Operation Iron Fist, was launched in 2002. Consistent with its past conduct, the LRA responded to Operation Iron Fist with a new campaign of violence against the civilian population with larger scale.

In March 2004, the Ugandan government launched another military offensive, Operation Iron Fist II, which included a renewal of the protocol with the Sudanese government. The LRA also responded with a number of massive attacks, most severely on civil targets. In the same year 2004, the Ugandan government started a new peace process, again led by Betty Bigombe. It appeared to have initial success, and presented the most hopeful prospects for a peaceful settlement to the conflict since 1994. A cease-fire with the LRA was secured at the end of November 2004, and subsequently extended a number of times until February 2005. By June 2005, the prospects for peace in the north remained less certain, and low-level fighting between the LRA and UPDF continued, as did LRA attacks on civilians. The failure of the Juba Peace brought on board forces of US, and South Sudan, in the hunt of the LRA by the Uganda forces. By 2013 the capacity of the LRA has been completely reduced and its remaining elements including its leader Joseph Kony were forced to the Democratic Republic of Congo and the Central Africa Republic where they are still hiding till now.

# Chapter 6

<hr/>

# War by Proxy

## Introduction

Like many colonial African countries, Sudan and Uganda contest their borders. This has created tension between them occasionally and they are only waiting for a right time to fight this border war. With civil wars in both countries, an opportunity for border issues to be capsuled within the package of other allegations was found. This made both LRA and SPLA factors in Uganda-Sudan relations.

## Uganda Grooming the SPLA

Sudan accuses Uganda of supporting its descendent - the SPLM/A. Uganda too, accuses Sudan of being behind LRA rebellion. As President Museveni explains:

*...by the time the NRA won victory in 1986, the SPLM was fighting on the Ethiopian border. There was not a single SPLM/A on the Uganda border. Why then, did Sudan arm the Erica Odwars, the Lakwenas, the Konys and others to attack us, starting with the 28th of August, 1986.*[190]

These accusations and counter accusations persisted till South Sudan broke away from Sudan on 9 July 2011, but mistrust still exist between Uganda and Sudan till today. The independence of South Sudan from Sudan ended the border conflict between Sudan and Uganda but transferred it to South Sudan. South Sudan is in civil war and is not prioritizing border issues now, even though a few skirmishes between the forces of the two countries were seen in 2020 along Kitgum Magawi districts. Small scale diplomatic efforts between the two countries took place along the borders but there has been no concrete outcome or solution to the conflict caused by disputed borders.

## Sudan Grooming the LRA

As already stated, the LRA is a factor in Uganda-Sudan relations. Between 1987-2005 tension between Sudan and Uganda developed sharply over accusation and counter accusation over the support of the rebels. Museveni personally believed that the LRA was an enemy groomed and supported by President el Bashir government[191], while Sudan accuses Uganda of providing SPLM/A with war logistics. The International Crisis Group, an interested party in the LRA war noted that: "...The Uganda army eventually forced Joseph Kony, the group's

---

190     Museveni, p. 262.
191     Museveni, p. 254.

mercurial leader, and his followers into Southern Sudan where they became for a long time a proxy force for the Khartoum government in the Sudanese civil war..." [192]

Mareike, a researcher on Sudan-Uganda relations, explains the confusion created by the LRA war between Uganda and Sudan as below:

> *The arrival of the Lord's Resistance Army (LRA) in Sudan in 1993-94 marked the beginning of more than a decade of fighting involving Ugandans on Sudanese soil. This development had an impact on both the Sudanese civil war and war in Uganda, isolating large parts of Sudan's Eastern Equatoria state from outside help and causing thousands to flee. The LRA had ventured into Sudan in the early 1990s to seek refuge from the fighting in Uganda. By 1993, the Sudanese government of Omer al Bashir had turned the LRA into a significant actor in Khartoum's efforts to crush the southern rebellion* [193]

In his study of the Horn of Africa Region, Samson Wassara notes that the Horn of Africa is host of multiple layers of conflict during the last years of the twentieth century[194]. Indeed, between 1980s- 2005, all five countries that are known as Horn Africa countries, with exception of Djibouti, were all in civil wars and were all in conflict over accusations and counter accusations over the support of rebels. The Uganda LRA civil war because of its transnational nature became regarded as one of the civil wars in the Horn of Africa.

---

192      Ibid.

193      Mareike, p. 10.

194      Samson Wassara. Why Conflict in South Sudan and Somalia is Beyond Prevention and Management. Africa Insight Vol 49(3), December 2019.

The LRA was conceptualized, organized, armed and used as a weapon by Sudan in response to its alleged Ugandan support to the SPLM/A which was battling the Sudan government by then. In a tit for tat, war Sudan had hoped to overthrow Museveni's government in Uganda and replace it with Khartoum's puppet government, as Museveni himself argued[195]. According to Museveni, Sudan after succeeding in bringing Joseph Kony into power hoped to join hands with LRA to fight and finish off the SPLM/A, which was fighting the Sudan government in the south[196]. This is political explanation behind Sudan's support for the LRA but there is a religious and cultural aspect as well. Since its coming into power on 30 June 1989 through a military coup, Al Bashir had advanced an Islamic agenda pursued by the National Islamic Front (NIF). The NIF has always viewed SPLM/A as an obstacle to the expansion of a wider mission to expand Islamic fundamentalism and Arab Culture in the region.[197]"

For these two reasons Sudan was obsessed to make sure that the LRA was in a better position militarily to achieve the task given to it by the Sudan. Unfortunately, the LRA loss of popularity among its own Acholi people because of its fighters' conduct, reinforced by military operations and pressure from the UPDF and the SPLA, weakened the LRA militarily and it was forced to relocate to Sudan where they continued to launch guerrilla warfare into Uganda. As Apuuli states:

*..,the LRA rebellion lost popularity among the people of northern Uganda because its waged war on its own people, the Acholi. reinforced*

---

195     See Museveni,p. 256.

196     Ibid.

197     Extract from anonymous document, 29 March 2002.

*by pressure from both the UPDF and Local Defence Units (LDUs),*
*the LRA and Kony fled to South Sudan, where he found fertile ground*
*to operate as the area has been wracked by war since the SPLA of John*
*Garang began fighting the Khartoum government in May 1983.*[198]

Indeed, as the phrase goes "the enemy of an enemy is a friend". On this principle, it was not a surprise that Sudan would welcome the LRA with no reservations especially when Museveni's sympathy with the SPLM/A was common knowledge. By 1987, Museveni had established that Sudan was behind LRA logistics, and responded by translating his existing sympathy for SPLM/A into full support. This support boosted SPLM/A moral and by early 1990s, the SPLM/A had made a big military gain against Sudan government forces stationed along the Uganda Sudan borders. In order to rescue itself from the pressure of the SPLA, Sudan resorted to using LRA as shield. Almost in every Sudan military deployment deployed in and around Juba suburbs, the external circle of its defensive positions became the LRA. This was followed with Sudan's massive military support in terms of ammunitions, weapons and food to the LRA forces. This was intended to give Museveni pressure to abandon its connection with SPLA in the South, unfortunately it made him adamant and instead, increased his support to the SPLM/A.

Al Bashir calculated that by providing military logistics to the LRA, the SPLA was struggling to survive after its fragmentation in 1991 especially when the SPLM break away groups surrendered to him and he turned them into his government militia, fighting alongside Sudan forces against SPLA. Generally, as a matter of fact, neither the Uganda

---

198     Kasaija Philip Apuuli, p. 54, also see Acker, p. 335.

nor Sudan government was innocent of supporting the other side's rebels. They both supported their adversaries objectively to maintain themselves in power but that became an advantage to the rebels, in particular the SPLA, who finally succeeded to get South Sudan's independence through the help of Museveni.

## Conclusion

The failure of Joseph Kony to sign the Juba Peace Talks mediated by South Sudan, reinforced Museveni's argument that the LRA is a lunatic organization which could only understand the language of gunfire. Indeed, Joseph Kony's failure to sign the deal shifted the nature of dealing with him and his organization from a peaceful to military solution. This led to launching aggressive military campaigns to hunt down Joseph Kony and the LRA, wherever they may be. Very significantly, this time the hunt was jointly executed with the involvement of US special forces, and South Sudan SPLA. This joint operation put the UPDF in an advantage position over the LRA. The mission was tasked to capture or killed Joseph Kony. Kony was not killed but his force was incapacitated and was forced to depart from Uganda. Kony's departure from Acholi land was a remarkable success for Museveni. Acholi people celebrated, and Museveni was rewarded with more votes by the Acholi people, in the 2011 presidential elections. As Izama and Wilkerson states:

> *Among the most startling aspects of the 2011 results was the rise of NRM support in northern Uganda...Long an opposition bastion, the North shifted toward the NRM with a force that heavily outweighed the slight increase in opposition support seen in the traditional NRM*

*strongholds of the West.*[199]

With this achievement, peace and economic growth returned to northern Uganda after the clearance of the LRA from northern Uganda, thus, bringing to the end LRA terrorism in northern Uganda which prevailed for couple of decades. At first, the LRA concentrated its battles on the government troops. However, as time went by it adopted a tactic of terrorizing civilians whom it claimed to be fighting for[200]. As a mechanism of protecting citizens against LRA attacks, the government introduced a policy to relocate about two million people from their homes in northern Uganda into protection camps. The policy was intended to deprive the LRA from recruiting in northern Uganda. With the signing of the CPA, northern Uganda region became stable, and Ugandan commercial goods began flowing across into South Sudan, and IDPs began leaving the misery of the camps in search of a better life.[201]

Significantly, although the LRA continue to launch raids into South Sudan from DRC, and from the CAR, its capacity of harming civil population within Uganda has ceased. However, regional and international observers remained critical about the LRA. In his briefing to the Security Council on the security situation in the Central Africa region, Abou Moussa pointed out that recent LRA attacks in South Sudan were a reminder to world leadership that the LRA remained a serious unpredictable potential threat[202]. Abu Moussa further called

---

199    Izama, A. and Wilkerson, M., 2011. Uganda: Museveni's triumph and weakness. Journal of Democracy, 22(3), pp.64-78.

200    Ibid.

201    Ibid.

202    Ibid.

for the UN regional strategy to address the LRA problem[203]. While the LRA continued to exploit every opportunity to regroup, the goal of eradicating the threat it poses is within the reach of the region.

By January 2013, a strategic development on how to handle LRA crisis in the eastern and central African countries, was undertaken. Highlighting such progress, Francisco Madeira, African Union special envoy for LRA Issues, recalled that in January of that year, key mission documents for the Regional Task Force had been adopted, including the standard operating procedures for the handling of persons suspected of involvement in LRA activities[204]. Indeed, in February 2013, the Congolese Armed Forces contributed manpower to join the Uganda and South Sudan forces in the regional fight against the LRA[205].

The appointment of the Mozambican envoy for the LRA as well as the Congolese acceptance to contribute forces to join the fight against the LRA confirmed AU commitment to resolving the LRA insurgency by all means [206]. The LRA had been pursued by Ugandan forces with participation of the SPLA South Sudan forces but unfortunately their efforts were constrained by funding and internal political dynamics which eventually brought civil war in South Sudan in 2013. Again, internal dynamics in the CAR resulting from the March *coup d'état*

---

203    See Security Council SC/11182. Briefing Security Council, Senior Envoy in Central Africa Calls for Sustained International Focus on Eliminating Lord's Resistance Army, Other Threats 20 November 2013available at https://www.un.org/press/en/2013/sc11182.doc.html

204    UN SC, 20 Nov. 2013 available at https://reliefweb.int/report/central-african-republic/briefing-security-council-senior-envoy-central-africa-calls

205    The strength of this force was 500 men.

206    https://sites.tufts.edu/wpf/files/2017/07/Lords-Resistance-Army-Mission.pdf

[207], not only delayed the participation of the Congolese forces, but also hindered the complete formation of the AU task force meant to pursue LRA from their hideout in the DRC, CAR and South Sudan[208]. The LRA exploited the delay in the execution of the regional force imposed by South Sudan and DRC. It allowed the LRA regroup and resume transnational cross-border movements shuttling between South Sudan, Western Equatoria State, CAR, and DRC. The result of these cross-border movements, was more displacement of civilians.

---

207    The Séléka rebel group overran the capital, Bangui, on 24 March, putting President François Bozizé to flight and naming Michel Djotodjia as the new head of state.

208    https://www.un.org/press/en/2013/sc11182.doc.htm

# Chapter 7

—◦◦◦◦◦◦◦◦◦◦◦◦◦◦◦◦◦◦◦—

# The Importance
# of Intelligence Sharing

## Introduction

In 2002, having confronted each other to a standstill over decades on allegations and counter allegation of aiding insurgencies, an intelligence sharing agreement between the Government of Sudan and Uganda to fight the LRA was successfully reached with the help of the Government of the United States. This agreement was meant to facilitate intelligence sharing about the LRA and other suspected terror organizations, such as Al Qaida, Boko Haram and so forth. The world increased interest in intelligence sharing - as necessitated by the threats that terrorism has posed on security of people and their property. This chapter presents the importance of intelligence sharing the fight against the ruthless LRA.

## Defining Intelligence

Defining intelligence is not as easy it is perceived, which is why 'intelligence' has been described as an umbrella term[209]. In fact, since World War II, much effort has gone into defining intelligence[210]. Intelligence has many facets to it. Intelligence covers a chain or cycle of linked activities from targeting, to the collection of data[211], which are either collected through human or signal intelligence. Therefore, intelligence is viewed as the plural of anecdote[212]. However, for purpose of this book, the term 'intelligence' is restricted to definitions as understood by Loch Johnson, Warner, and the CIA. Loch defined intelligence as "the knowledge, ideally, foreknowledge sought by nations in response to external threats and to protect their vital interests, especially the wellbeing of their own people[213]. According to Warner, intelligence is secret, state activity to understand or influence foreign entities[214]. And according to the Central Intelligence Agency (CIA), intelligence covers all military, economic, political, scientific, technological, and other aspects of foreign developments that pose actual or potential threats to US national interests.[215] Loch, Warner and the CIA provided helpful

---

209     See Peter Grill and Mark Phythian. 2006. Intelligence in an Insecure World. Polity. P.2.

210     Philip Davies. Ideas of Intelligence: Divergent National Concepts and Institutions. In Andrew, C., Aldrich, R. J., & Wark, W. K. (Eds.). (2009). Secret intelligence: A reader. Routledge. P. 13

211     Peter Grill and Mark Phythian. 2006. Intelligence in an Insecure World. Polity. P.2.

212     Moynihan, D. P. Secrecy, Yale University Press, p.202.

213     L.K. Johnson, 'Intelligence', in B.W. Jentleson and T.G. Paterson (eds), Encyclopedia of US Foreign Relations. Oxford University Press, 1997, pp 365-73.

214     Warner, M. 'wanted: A Definition of Intelligence'. Studies in Intelligence, 46(3), 2002.

215     Central Intelligence Agency: A Consumer's Guide to Intelligence, p.42.

definitions for our understanding of intelligence.

While it has been difficult to give a specific definition of intelligence, as Stephen Marrin clearly states: "Intelligence means many things to many peoples and boiling it down to a single definition is difficult,[216] nevertheless many scholars and institutions were generous to give us working definitions. In this regards, the Canada's intelligence community consists of a complex web of functionally differentiated agencies for the collection, assessment and protection of security-relevant knowledge on behalf of this country's foreign policy, security and defence establishment."[217] Warner applied Hoover Commission of 1995 and defined intelligence like this: "Intelligence deals with all the things which could be known in advance of initiating a course of action". Carl and Bancroft define intelligence as the product resulting from the collecting and processing of information concerning the activities and potential situations relating to domestic and foreign activities and to domestic foreign or U.S in an enemy held area. The Joint Intelligence provided another definition with military undertone as: Intelligence is the product resulting from collecting, processing, integrating, evaluation, analysis and interpretation of available in information concerning foreign nations, hostile or potential hostile forces of element or areas of actual or potential operation[218]. Another definition from Alex and Schmid states that: "Intelligence is a knowledge resulting from

216    Marrin, S., 2004. Preventing intelligence failures by learning from the past. International Journal of Intelligence and CounterIntelligence, 17(4), pp.655-672.

217    Martin Rudner Contemporary Threats, Future Tasks: Canadian intelligence and the Challenges of Global Security in Norman Hillmer & Maureen Appel Molot, eds., Canada Among Nations 2002: A Fading Power (Toronto: Oxford University Press, 2002, pp.141-171.

218    Joint Chief of staff 2009 GH:11

detection, collection, integration, evaluation, analysis and interpretation of information used for decision making for diplomats military and other operations".[219] From these few provided definitions we can conclude that intelligence is that refined part of information collected which is used to help leadership make informed decisions.

As we entered the 21st Century, there is still no widely established definition of intelligence, instead each author draws upon his/her definition based on personal experience[220]. Mark Lowenthal, in his textbook on intelligence and policy, suggests that the word 'intelligence' can be analysed in three different ways. First, as a process, of collecting, analysing and delivering information to the 'consumers', such as policy makers or operational commanders. This is often called as the 'intelligence cycle', although the essence of this cycle is now widely controversial[221]. Second, as a product, formerly circulated as paper, but now distributed through multilevel secure electronic databases. Finally, we can talk of intelligence services and intelligence communities as institutions.

As their name suggests, they often deliver various 'services' to the government and increasingly this involves active efforts to shape the world as well as merely reporting about it[222]. It is now obvious that the task to define 'intelligence' and its importance is not an easy one; therefore, the final notion will be presented at the end of this book. It can be clearly understood from them that intelligence is a box of

---

219    Alex and Schmid (2000).

220    Warner, M. (2009). Wanted: A definition of 'intelligence'. in C. Andrew, R. J. Aldrich, & W. K. Wark (Eds.), Secret Intelligence: A Reader (pp. 3-12). London: Routledge.

221    Hulnick, 2006.

222    Lowenthal, 2002, p. 8.

interrelated items that go beyond physical security to include economics, politics, technology, and most importantly about follow up of foreign nations activities. Also, these definitions as go far as to include cycle of planning, direction, collection, processing, analysis, production and dissemination. Indeed, intelligence collected on the LRA had involved planning, direction, collection, processing, analysis, production and dissemination of information among the parties concerned.

## Purpose of Intelligence

The purpose of intelligence is as clear as the sky. It is needed for security reasons. The purpose of intelligence is never static. It depends on what, where, when and how it is needed. Indeed, the last couple of years, organizations have demonstrated an increased willingness to exchange information and knowledge regarding vulnerabilities coming from terrorists' threats, and from other evil organizations such as LRA. This requires strategies, in order to collectively protect against today's attacks such as those launched by the LRA. The purpose of intelligence collection and sharing is to bestow a relative security advantage to avoid military and terrorist surprise attack.

## Methods of Intelligence Gathering

According to Peter Gill, the methods used to acquire information are placed into four broad categories[223]. These categories include open-source materials. Open-source material by itself is comprised of a publicly available data, speeches, official documents, newspapers, magazines among many others of such kind. In 1970s, Harry Howe Ransom estimated that about 80 percent of intelligence materials

223      Gill, p. 62.

specially in peace time are overtly collected from non-secret sources such as newspapers, libraries, radio broadcasts, business and industrial reports, or from accredited foreign service officials[224]. The remaining bulk of categories of information are limited and secret in nature, such kind of information are kept as secret[225]. Intelligence can be gathered either through Human Intelligence (HUMINT), Signal Intelligence (SIGINT), or through Imaginary Intelligence. Human intelligence is the oldest method used in information collection. It is the collection of information from human sources. This include using methods such as witness interviews or espionage. Signal Intelligence is one of the methods used to collect intelligence. Signal intelligence is also known as electronic intelligence. The electronic intelligence is collected by any form including by ships, planes, satellites. Also, interception of communications between parties is classified as signal intelligence. The Imaginary intelligence basically consists of photo intelligence.

During the Cold War era, intelligence sharing among blocks members increased rapidly. This saw Western Europe developing new methods of collecting intelligence, thus prompting development of new technologies, resulting by the 1980s, in a substantial improvement in the use of satellites[226], that became Signal Intelligence (SIGINT). That development in technologies was not without challenges among the participants. However, the flow of intelligence had been much smoother between the US and the UK especially in Signal Intelligence

---

224    Ransom, H.H. The Intelligence Establishment, Cambridge, MA, Harvard University Press, p. 81.

225    Ibid.

226    See Adriana N. Seagle, p.558.

(SIGINT)[227]. This was due to growing mistrust between the NATO members. As Seagle notes:

> *In the Cold War period, the relationship between Paris and London was marked by mistrust as de Gaulle consistently viewed Britain as an American asset in NATO.*[228]

Following the end of the devastating Second World War the world most powerful nations namely the US, the UK, France, and Russia saw the logic for the need to avoid such similar World Wars and catastrophes. The result was the emergence of an intelligence sharing approach among these great powers. Perhaps the need to share intelligence was influenced by the Realists school which dominated international politics after the end of the World War II. The Realist school argues that by nature, humans are motivated to seek domination over others, making politics among nations a struggle for power, and realpolitik policies the necessary prescription for survival[229]. In this sense, it is argued that the popularity to share intelligence is derived probably from its envisaged contribution to the maintenance of peace and security. After the World War I and the World War II the need to increase intelligence became apparent, because the nature of destruction in terms of human lives and property was far beyond normal. Indeed, UN was formed to maintain international peace and security. Furthermore, with advent of War on Terror, intelligence sharing even gained more importance.

---

227     Ibid.

228     Ibid.

229     Allison, G.T. and Zelikow, P., 1971. Essence of decision: Explaining the Cuban missile crisis (Vol. 327, No. 729.1). Boston: Little, Brown.

## Intelligence Sharing

What is intelligence sharing? As its name obviously shows, intelligence sharing is the sharing of intelligence information between agencies within the same government machinery or with other governments machineries. It is the ability to exchange intelligence, information, data, or knowledge between parties who are guided by common interest, for example between the US, Uganda and South Sudan that is meant to facilitate the use of actionable intelligence to a broader range of decision-makers.

Intelligence sharing between US, Uganda and Sudan began with the events of 9/11. Prior to 9/11 there were no such open arrangements concerning intelligence sharing between the US, Uganda, and Sudan. The US, though directly not affected by the LRA conflict, became interested in the fight against LRA. The LRA had been designated a terror organization by Obama administration. This interest was driven by the fact that the LRA was labelled a terror unit. The US desired to prevent further terrorist strike on its soil especially after the 9/11 events. The US had other interest to achieve through its involvement in the fight against LRA.

By bringing bitter enemies Sudan and Uganda together, the US found an opportunity to improve Sudan and Uganda relations and use these improved relations to help push for the peace agreement between southern and northern Sudan[230]. Uganda and South Sudan were directly affected by the LRA war. The LRA killed, abducted and displaced hundreds of people along the Uganda-South Sudan border. Sudan for its own part, wanted to improve relations with the US amid terrorism accusations that Sudan remained a state sponsor of

---

230     See Hilde F. Johnson. P. 18.

terrorism[231]. Thus, the Sudanese government not only had more reason to placate the Americans, it also had something to offer.[232]

However, it has been and is yet still unclear under the intelligence sharing agreement among these countries, what type of information is to be shared, since a country can decide what information to share and what to keep for its own use. We know that various agreements signed on intelligence sharing allow for the exchange of specific types of information. Usually, information on passengers or travellers, migrants and immigrants or citizens, are usually exchanged most the time without the knowledge of the persons on whom this information is gathered. This has generated a public debate around intelligence-sharing, with emphasis on the dangers of interfering with individual privacy - as far as the gathering and storing of identification data for further use may represent[233]. Yet, the problem seems to go much deeper than that. However, intelligence-sharing has also resulted in the promotion and cultivation of good relationships between agents of diverse security organizations.

With regards to intelligence sharing on the LRA, the underlining common factor has been centred on gain. Each of the four countries - Uganda, Sudan, South Sudan and US participating in the LRA fight, had something to gain, unfortunately it has been dimmed and it is unclear what type of information they were sharing between them. For instance, did the sharing agreement cover information on Joseph Kony specifically or did it include the LRA top commanders or all LRA members? What about information on the alleged support to

---

231     Ibid.

232     Ibid.

233     See American Civil Liberties Union 2003.

LRA by Sudan or Uganda to SPLA? Did sharing go beyond LRA to include Sudan's involvement with terror organization, which has been the main interest of the US?

All these questions need to be answered within the context of national and international security aa intelligence to be collected is significant to security at every stage. As Gill argued, intelligence is significant for both domestic and international security[234]. Indeed, intelligence sharing is vital for survival of a state considering the nature of international politics and its anarchic nature. The nature of international politics is that countries interact pursuance to their national interests. Thus, the interaction among and between US, Uganda, Sudan and South Sudan is based on such model. They interact purposely in pursuant and fulfilment of their national interests. US wanted its nationals secured from terrorists' attacks and as well to protect them from human rights abuses, which is a major value of the foundation of the US at independence on 4 July 1776.

Equally, Uganda and South Sudan have responsibilities of protecting the lives of their people who the LRA have been threatening for so long. For its own part, Sudan's main interest in joining intelligence sharing arrangement against its former ally the LRA has been to secure US removal of its name from the list of states sponsoring terrorism. Earlier, in one of the Security Council meetings, the US indicated its willingness for Sudan to be removed from the list of those states sponsoring terrorism if it collaborates fully in the War against Terror. The terror attacks of 9/11 complicated Sudan's position[235].

---

234    Gill, p. 172.
235    Hilde, p. 18.

Realizing importance of intelligence sharing, the United Nations Department of Peacekeeping Operations (DPKO) adopted in 2006 a policy that a Joint Mission Analysis Centre (JMAC) and a Joint Operations Centre (JOC) be established in all PKO to conduct all-source intelligence branch of the UN force[236]. Such operations improved enormously the capacity of the United Nations to meet some of its most challenging mandates[237]. This is thanks to the UN mission in Haiti, which recorded intelligence sharing in the UN operations system for the first-time in the twenty-first century[238]. The United Nations Interim Administration Mission in Kosovo (UNMIK) was another twenty-first century mission which pioneered intelligence-led operations, especially to deter, target or capture the 'spoilers' of the peace process and criminal elements[239]. Between 2006 and 2007, it was believed that such an approach allowed the mission to gain ascendancy over gangs who controlled large sections of several Haitian cities, particularly the capital Port-au-Prince[240].

The JMAC had been created in 2005, at the urging of the UN Security Council, as an integrated unit of military officers, police and expert civilians. Its mission was to collect information and to produce useful intelligence to help the mission leadership to make informed

---

236    A. Walter Dorn. Intelligence-led Peacekeeping: The United Nations Stabilization Mission in Haiti 2006–07. Intelligence and National Security. 24, No. 6, 805–835, December 2009.

237    Ibid.

238    ibid

239    See Ben Lovelock, 'Securing a Viable Peace: Defeating Militant Extremists – Fourth-Generation Peace Implementation' in Jock Covey, Michael J. Dziedzic and Leonard R. Hawley (eds.), The Quest for Viable Peace: International Intervention and Strategies for Conflict Transformation (Washington: US Institute of Peace Press 2005) pp.139–40, 144.

240    Ibid.

decisions. Central to this was the reliance of JMAC on the information collected through the use of local informants to determine the locations and activities of gang[241]. The vitality of intelligence sharing was further noted in other United Nations missions. In the Democratic Republic of the Congo (MONUC), for example, the regional Eastern Division headquarters in 2006 was given control over the movements of soldiers in the field tasked to obtain information about dangerous rebel groups hiding in the jungle. This obviously suggests that sharing of intelligence has not been confined to terror issues; it ranges from issues as diverse as terrorism, human trafficking, insurgency, war, poverty, and unemployment or daily issues that can affect national security adversely.

## Intelligence Cycle

Mark Phythian addresses very sufficiently the intelligence cycle in his book *Understanding the Intelligence Cycle*[242]. As he states, the intelligence cycle involves a process of developing raw information material into a refined intelligence that is intended to be relied on by the policy makers. The intelligence cycle composes of issuance of requirements by policy designers, collection, processing, analysis and evaluation and of course dissemination of intelligence. Since World War II, much effort has gone into defining intelligence and the intelligence cycle[243]. This search for definition validated the importance of intelligence in our today world and throughout the history of humankind, and it

---

241      Ibid.

242      Phythian, M. (Ed.). (2013). Understanding the intelligence cycle. Routledge.

243      Andrew, C., Aldrich, R.J. and Wark, W.K. eds., 2009. Secret intelligence: A reader. Routledge. p.13

has been realized that intelligence sharing, which is a processes of intelligence cycling with foreign counterparts, offer sa number of benefits that has proved to outweigh risks. Intelligence sharing and cycling gives indications and warnings of an imminent attack[244]. In this way, it is argued that intelligence cycling is vital for its production of refined information that are in turn, used to safeguard the UN mandate of international safety.

The terrorist attacks in the United States on 11 September 2001 forcefully brought to the fore the necessity for the scrutiny of intelligence gathered through cooperation among security and intelligence agencies at the national and international levels[245]. The intelligence cycle is intended to bring concrete results about the terrorists and their organizations. The US has always prioritized intelligence collaboration with foreign nations in Africa, America, Asia, and Europe. As the US congressional research service reports: "From its inception, the United States Intelligence service has relied on close relations with foreign partners"[246]. Rosenbach and Peritz also echoed this view and noted that U.S. national security interests have long rested upon international cooperation between intelligence services.[247]

Mark Lowenthal advises that intelligence should be looked in three

---

244     Congressional Research Service, 15 May 2019. United States Foreign Intelligence Relationships: Background, Policy and Legal Authorities, Risks, Benefits. Congressional Research Service https://crsreports.congress.gov

245     Lefebvre, S., 2003. The difficulties and dilemmas of international intelligence cooperation. *International Journal of Intelligence and Counterintelligence, 16*(4), pp.527-542.

246     United States Foreign Intelligence Relationships: Background, Policy and Legal Authorities, Risks, Benefits. Congressional Research Service https://crsreports.congress.gov R45720, May 15, 2019.

247     Rosenbach, E & Peritz, A.J. Intelligence and International Cooperation. *Belfer Center for Science and International Affairs, Harvard Kennedy School, July 2009. Available at* https://www.belfercenter.org/publication/confrontation-or-collaboration-congress-and-intelligence-community

ways[248]. First as an intelligence cycle, a process, through which intelligence is requested by policymakers or operational commanders, then collected, analysed and fed to the consumers. Second, intelligence is defined as product, once circulated as paper, but now increasingly distributed through multi-level secure electronic databases. And thirdly, Mark Lowenthal suggests that intelligence can be viewed as service and community of institutions. Michael Herman describes intelligence as a form of state power just like economic and military powers which are usually associated with great powers[249]. Sun Tzu gives definition of intelligence in the following words: "He who knows himself, knows his friends and knows his enemy will be successful in a hundred battles".[250]

Often, intelligence relationships reflect mutual security interests and the trust that each partner has of the other's credibility and professionalism.[251] Credibility and professionalism within the context of US intelligence service are generally strategic and cover a range of national security priorities which is involving national defence, emerging threats, counterterrorism, counter-proliferation, treaty compliance, cybersecurity, economic and financial security, counter-narcotics, and piracy. US intelligence relations with foreign friends offers a number

---

248     Mark Lowenthal in Andrew, C., Aldrich, R.J. and Wark, W.K. eds., 2009. Secret intelligence: A reader. Routledge. P.1.

249     Herman, M., 1996. Intelligence power in peace and war. Cambridge University Press.

250     Sun Tzu quoted in Herman, M., 1996. Intelligence power in peace and war. Cambridge University Press. p. 1.

251     Rosenbach, E & Peritz, A.J. Intelligence and International Cooperation. Belfer Center for Science and International Affairs, Harvard Kennedy School, July 2009. Available at https://www.belfercenter.org/publication/confrontation-or-collaboration-congress-and-intelligence-community

of benefits. These include indications or giving warning of an attack, as well expanded geographic coverage, corroboration of national sources, accelerated access to a contingency area, and a diplomatic backchannel. However, it also presents risks of compromise due to poor security, espionage, geopolitical turmoil, manipulation to influence policy, incomplete vetting of foreign sources, over-reliance on a foreign partner's intelligence capabilities, and concern over a partner's potentially illegal or unethical statecraft.

As the intelligence world practitioners continue to develop new strategies of information-sharing relationships with foreign entities, it is important that parliament members should continue to have oversight of the intelligence. This is because foreign intelligence services are mostly concerned with their own self-interest. However, liaison relationships between intelligence services can sometimes allow for warmer partnerships between the countries. For example, the US and China maintain a military-to-military relationship despite existence of contentious political history because the US generally views the partnership as a mechanism to minimize miscalculations between the armed forces.

Cooperation between intelligence agencies is essential for maintaining national and regional security. The Sri Lanka Defence Secretary Gotabaya Rajapaksa notes the significance of intelligence and writes: "Having a proper understanding between the Financial Intelligence Units, Intelligence Agencies, Law Enforcement Agencies within the region is therefore extremely important in tracking illegal financial transactions and in identifying and apprehending the culprits involved".[252]

---

252    Gotabaya Rajapaksa. "Importance of Intelligence Sharing Among Nations" Asian Tribune, Wed, 2013-05-08.

The most important tool nations have at their disposal in countering these threats, is intelligence sharing. For intelligence to be more effective, however, it needs to be shared amongst nations. It is also very important to realize that few of the serious threats nations face in today's environment are truly localised. Terrorism, human trafficking, narcotic drug smuggling and the illegal financial transactions which support all these activities, are not confined to one nation but take place within several nations and across national borders. For instance, when the Liberation Tigers of Tamil Eelam (LTTE) was engaging in ruthless terrorism activities in Sri Lanka, it raised funds abroad and used agents in various parts of the world to smuggle arms, ammunition and heavy weaponry to Sri Lanka by sea. As a Human Rights Report mentions:

> *"Even after the end of the war, the rump of the LTTE organization is still engaged in raising funds and trying to regroup in order to destabilize Sri Lanka through various means. These elements have even gone to the extent of providing arms training abroad to LTTE cadres in recent times. Because of the effective cooperation between Intelligence agencies, however, we were able to identify and stop these developments in their early stages. This is an example of how enhanced cooperation between Intelligence agencies is essential for maintaining national and regional security"*[253].

Improper intelligence cycling leads to intelligence failure. In 1941, a

---

253     Human Rights Watch, Sri Lanka: Funding the "Final War": LTTE Intimidation and Extortion in the Tamil Diaspora, 15 March 2006, C1801, available at: https://www.refworld.org/docid/4455d48f4.html [accessed 24 April 2021]

deep penetration mission took Russian intelligence by surprise. Hitler staged an operation invading Russia with 3 million German troops who poured in from the Arctic Circle to the Black Sea. The Russians had plenty of information about troop movements eastward by the Germans and couldn't help but notice the increased number of aerial surveillance flights Hitler was sending over Russia. On the other hand, US intelligence had already told Russian intelligence of Hitler's plans to invade Russia back in 1940. Russia was convinced that similar intelligence leaked to them by the British was really counterintelligence. Hitler played two deception schemes. He first explained the build-up of his troops on the Russian border as being there for training purposes, to prepare for the invasion of England. He then explained them as being contingency forces against possible hostile Soviet action. Stalin bought all this because his own intelligence led him to believe that Hitler would not dare try to fight a war on two fronts.

Similarly in 1941, a task force of 33 Japanese ships stationed themselves 200 miles north of Oahu and launched two successive waves of air surprise attack on Pearl harbour[254]. By the time the attack was over, the U.S. had lost 18 warships, 200 airplanes, and over 2000 personnel. The case of Pearl Harbor is regarded as the worst case of intelligence failure in history. No intelligence agency had prepared a report for the possibility of an attack there, although everyone talked about it. Naval intelligence did not even have a minimal amount of strategic or tactical intelligence. They thought Japan would attack Thailand about that time of year. The problem was that America lacked Human Intelligence (HUMINT) on Japan. The U.S. had a few geisha girls on the payroll, but no agents in the Japanese elite. The U.S. had broken

---

254     It was believed that about 350 Japanese war planes participated in the attack.

the Japanese code, but what they were intercepting was just diplomatic and espionage information (movement of spies), nothing of the nature of military plans, and the Japanese also they changed their codes a day before the attack. Japanese radio transmissions deceived the Americans into thinking the task force was assembling for training manoeuvres.

It is believed that from 1998-2001, Osama bin Laden's terrorist network, al-Qaida, changed their new operatives to the US from Hamburg, Germany and Bangkok, Thailand while the CIA was watching and disrupting other sleeper cells. Although intelligence officials had long speculated on the use of airliners as weapons and knew that Bin Laden was a potent adversary – as demonstrated by the 1998 attacks on Dar es Salaam and Nairobi embassies and a 2000 attack on the US Cole, the infiltrators managed to avoid attention by getting six different kinds of fake ID. Some even attended pilot training, a fact the FBI picked up on as early as 1998 and something British intelligence tipped the US about in 1999, which led to a 1999 US intelligence brief, which political officials claimed they never heard of or claimed was vague.

Another similar intelligence was forwarded in 2000, when the US conducted a "Dark Winter" drill in which fictional terrorists flew fictional planes into buildings. In early 2001, a flight school alerted the FAA about a suspicious student, and the summer of 2001, several suspected terrorists were seen drinking and partying in Las Vegas. Also, in late summer 2001, a Phoenix FBI agent sent a warning memo to his supervisors, and Russian, Jordanian, British, and Israeli intelligence all tried to warn the U.S. In August of 2001, a senior FBI counterterrorist official quit. The Minnesota FBI began working with the CIA on detained suspect Moussaoui, and the CIA issued another intelligence brief to the President. On September 10, 2001, it was alleged that a group of top Pentagon officials suddenly cancelled commercial flight plans, financial centres reported a surge in money transfers from banks in the World Trade Centre. While these scenarios may offer us something, it also suggests to us that having an information is one thing and utilizing the information is another thing.

## Intelligence Sharing and Security Studies

The Realist theory of international relations understands intelligence sharing within the context of security studies. This theory is generally considered the most coherent interpretation available of the nature and causes of political events, ranging from armed struggles, military alliances, diplomatic negotiations to relations between states[255]. Traditional realists describe the interaction between states in terms of enmity and collaboration. But neo realists like Keohane and Nye consider states to be sensitive and vulnerable as well[256]. Waltz presents survival instinct of states[257].

The Realist theoreticians suggest that states' behaviours are motivated by national interests – particularly economic but also increasingly driven by perceived national security threats, such as the flow of refugees or radicalism. Indeed, it is the radicalism of the LRA that

---

255     Palan, R. and Blair, B. On the idealist Origins of the Realist theory of international relations. In Little, R. (1993). Eds. Review of International Studies. Volume 19, Number4 – October 1993, Cambridge University Press, p. 385.

256     See Wolfers, A. (1962). Discord and Collaboration: Essays on International Politics. Baltimore.

257     See Keohane R. and Nye, J. (1979). Power and Interdependence. Boston Theory of International Politics. Reading: Mass respectively.

united US, South Sudan with Uganda. However, several analysts have noted the wide gap between the capacity required to gather and share intelligence and the capacity available. Koerner and Gebrehiwot point out that building institutional capacity does not only entail recruiting more staff and providing more funding: "It also requires increasing the efficiency and impact of extant organizational arrangements through organizational improvement."[258]

Governments collect, process and use information for positive purposes not for detriment. This is the rationale behind intelligence sharing. Organized intelligence sharing increased in the 20th Century period and since then, most governments have institutionalized it.[259] To understand why and how the process of intelligence sharing came about, it is necessary to look at two subject areas: firstly, the evolution and development of the conceptual fields of security studies; and secondly, the actual process of intelligence sharing between USA, Uganda, South Sudan and Sudan.

One of the issues studied within the field of politics is security, which at first, was mostly tackled within strategic studies where war study is a subfield. Strategic study is a policy-oriented field, which focuses on issues of security from a national perspective. For a while this served its purpose; the dominating states maintained power and focused only on security as a national issue and only on their own security. These are clear in the works of Roberta Wohlstetters, Victor

---

258    Koerner, M and M Gebrehiwot (2013) 'Capacity building for APSA: dos and don'ts in programme design' in Engel, U and J Porto (eds) Towards an African Peace and Security Regime. Surrey: Ashgate.

259    Michael Herman (1996). Intelligence Power in Peace and War. Cambridge University Press. p. 2.

Marchetti and John Marks, Christopher Andrew, Uri Bar-Joseph, Bruce Berkowitz, Harold Ford, Lawrence Freedman, Roger George, Roy Godson, Alan Goodman, Glen Hastedit, James I. Walsh, Michael Herman and many others.

Intellectuals wrote number of books when they were seeking ways to understand what had caused such horrific events such as the World Wars and looking for ways of preventing them from happening again. One of these ways, has been the design of the intelligence sharing mechanism. In his book, *The International Politics of Intelligence Sharing*, James Igoe Walsh (2009) presents strategies for securing more reliable intelligence. He argues that:

> *Countries that seek information get concern of other state's intelligence efforts. James continues to argue that 'states regularly draw agreement in which one power directly monitors and acts on another power's information-gathering activities-a more streamlined approach that prevents the dissemination of false 'secret'.*[260]

In developing this strategy, Walsh draws on recent theories on international cooperation and evaluates both historical and contemporary case studies of intelligence sharing. In his article "Intelligence sharing in the European Union: Institutions Are Not Enough", James I. Walsh revealed that the European Union (EU) has developed three institutions to facilitate intelligence- sharing between its Member States: The Berne Group, Europol and the European Union Military Staff.

According to him, these institutions serve the useful function of creating technical mechanisms for the diffusion of intelligence among

---

260      James I. Walsh (2009), The International Politics of Intelligence Sharing.

national authorities. He however, noted that: *"these institutions do not tackle the problem of mistrust, which is the key barrier to fully effective intelligence- sharing*[261]*"* Richards K. Betts, views intelligence as power. He argues that power without knowledge is useless and dangerous.[262]" Peter Gill and Mark Phythian in their book *Intelligence in An Unsecure World* (2006) defines intelligence in terms of security. They argue that: "intelligence is an inherently competitive pursuit with a purpose to achieve a relative a security advantage."[263]

Furthermore, they maintain that "since intelligence is about the production of knowledge, with agencies, operating at the cutting edge of deploying new information and communication technology, it is entirely appropriate to explain it with an approach to social theory that itself emphasizes the significance of new information technologies in reshaping and subverting modernist methods of generating knowledge."[264]

In this article, "International Intelligence Co-operation", Stephen Lander argues that intelligence services and intelligence collection are at the heart of the manifestations of individual state power and of national self- interest[265]. In Michael Herman's book *Intelligence Power in Peace and War*, it is argued that intelligence is drawn on for any kind of national action, including intervention in others' conflicts or

---

261    James I. Walsh Intelligence-Sharing in the European Union: Institutions Are Not Enough, JCMS 2006 Volume 44. Number 3. Pp. 625-43.

262    Betts, Richard K. (2007), 'Enemies of Intelligence: Knowledge and Power' in p.1.

263    Peter Gill & Mark Phytian (2006), Intelligence in an Insecure World. P. 1.

264    ibid, p. 23.

265    Stephen Lander in Christpher Andrew et al (2009), Andrew, Christopher, Richard J. Aldrich and Wesley K. Wark (2009), Secret Intelligence: A reader p. 140.

mediation between them with international security motives[266].

In the 1950s John Herz coined the phrase 'Security Dilemma,"[267] to refer to the situation whereby one state's focus on increasing its own security could lead to greater insecurity for others as they asses their own position and feel threatened. The dilemma can come about regardless of whether or not a state intends to make others feel insecure and the action of generating greater international insecurity by supposedly creating greater national security began to be regarded by many as a paradox. This paradox is evident on all levels, from the macroscopic level whereby one country's ownership of nuclear arms makes another country feel it can be secure if it too, acquires the same, to the microscopic level dealt with in this study, whereby intelligence sharing may make an individual state feel safer. As Michael Herman puts it: "States want accurate information and good forecasts about other states to help them formulate intelligence defences', it has become indeed important to look at them."[268]

In the 1980s, academics working on the concept of security began to look to a more collective, global approach to the issue of security. Leonard Beaton, L.B Krause, Joseph Nye, Hedley Bull, and the Brandt Commission all called for a new approach, which acknowledged security as a subject most usefully, addressed multilaterally. As Booth states: "Those strategies who do not attempt to be part of the solution will undoubtedly become an increasingly important part of

---

266    Michael Herman (1996), Intelligence in Peace and War, p. 156.

267    Hertz, J. Idealist Internationalism and the security dilemma. World politics, January 1950, 20, no.2. pp157-80.

268    Michael Herman, p. 165.

the problem."[269]

Part of the difficulty of tackling the subject of security has been the lack of an agreed definition of security. In 1983, in his work *People, States, and Fear*, Barry Buzan examined the reasons that he believed security was an underexplored concept. He disagreed with the notion that security is no more abstract than other concepts such as power, peace, equality or freedom. He insisted that the discussions must move beyond the field on national security and encompass a broader range with the field of security studies operating somewhere between the power focus of realists and the peace focus of idealist; a discipline that would include the political, military, economic environment and societal elements of security. In this book, Buzan looked at the state has the 'irreducible base object, however according to Steve Smith, in later years, Buzan was struggle to maintain this position as more academics argued that the individual should be the referent object for security,[270]" This debate is at the core of decision making with regard to security, as Robert Cox observed: "all theories are created for some purpose."[271]

Although security studies attempt to explore and challenge the security dilemma, the Realist school of thought remained dominant. As Booth explained:

*Realists derived security studies continues to survive and flourish*

---

269    Booth, K. Strategy and ethocentricism. London, Croom Helm, 1979.

270    Smith, S.'The Contested Concept of security.' In Booth, K. (ed), Critical Security Studies and world politics. Colorado, Lynne Rienner Publishers, 2005., p.33.

271    Cox, R. 'Social forces, states and World order.' Millennium: Journal of international Studies, 1981, Vol. 10. No. 2. Pp. 126-155.

*because the approach is congenial for those who prosper from the intel-*
*lectual hegemony of a top-down, statist, power centric, masculinized,*
*ethnocentric, and militarized worldview of security.*[272]

Booth and others invented the term 'critical studies' in 1994, making the argument that by searching the solutions within the system,
the Realist approach perpetuates the insecurity of the majority of people worldwide. Booth views realist elitist and traditional which tries
to view contemporary issues with superficial and unworkable mind[273].
Smith maintains that realists redefine existing order, treats the observed relationship as unproblematic, reports neutrally on the existing natural order, works
within the scientific paradigm of positivism.[274]

The two postulated that instead of aiming for objectivity, students
of security studies should work towards approaching the subject from
critical distance. Vital to Jones and Booth argument that security studies should be focused on human emancipation.

At this stage, it is also necessary to look back at the intelligence
sharing between Sudan and Uganda. Unlike in other cases, Sudan
and Uganda through US intervention were busy playing games which
look to be promoting and strengthening intelligence sharing between them. In 1993, though, the United States linked a car bomb
at the World Trade Centre in New York to Osama bin Laden, a Saudi
Islamic fundamentalist living in Sudan. Khartoum was then added to

---

272     Booth, K. (ed.), Critical Security studies and world politics. Colorado, Lynne
Rinner Publishers, 2005, p.9.

273     Ibid, p.4.

274     Smith, S. op. cit.p.42.

the State Department list of state sponsors of terrorism[275]. This was an opportunity for the Government of Uganda to effectively engage Sudan. The Government of Uganda (GoU) had to seriously engage the Government of Sudan and its Army (SAF) in practical ways aimed at addressing the LRA insurgency through a cooperation protocol with UPDF. As a result, the GoU put forward three options to the GoS. Firstly, it gave Sudan and its army an option to fight the LRA by itself. Second option is to do jointly with the UPDF. And the third option specifically requested SAF to allow UPDF to do it.

Sudan, in order to show commitment on its part, responded by inviting UPDF to enter into its territory to pursue LRA rebels stationed inside Sudan. This is the first time, government of Sudan (GoS) has openly invited foreign forces to carry out military operation inside Sudan. As part of a show of good will, the GoS accepted members of the Ugandan External Security Organization (ESO) to be stationed in their operation command basis in Juba, Torit and other locations inside southern Sudan. Joint mentoring teams were instituted.

All these attempts to address the question of terror were significant in influencing the US and the AU to joining in in the fight against the LRA. Through this arrangement much has been achieved and the LRA has been de-toothed and is currently operating in hiding, reported at times in DRC, CAR and Sudan. This was also a big relief to the Uganda government which fought the Lord's Resistance Army in an attempt to protect the people of northern Uganda and to also save its image from failing to defeat the LRA, which waged war on the Ugandan government and its own people the Acholi in northern

---

275    Rebecca Hamilton. How America was sold on South Sudan. Reuters special Report 1. Washington, July 11, 2012.p.6.

Uganda with unbelievable suffering.

The long social disorder that the National Resistance Movement of President Museveni, inherited in 1986 from previous governments and especially following the downfall of government after government, ending at last with downfall of Acholi-led Okello regime, contained the root causes for continue insurgency. This social disorder was amplified by external circumstances of the Cold War that were breeding ground for rebellion almost all over Africa. Also, it has been noted that the Acholi region remained to be living within poverty line, in terms of any basic human development indicator[276]. Further, it has been explained that the Acholi community voted overwhelmingly against the National Resistance Movement, in local, parliamentary and presidential elections, thus being viewed as anti-Museveni.

Perhaps the longest lasting intelligence collaboration is with the "Five Eyes" group: comprised of the United States, the United Kingdom, Canada, Australia, and New Zealand. The alliance developed out of World War II era agreements between the UK and US to share signals intelligence and has evolved into a broader undertaking. In 1955, the arrangement was extended to Australia, Canada, and New Zealand. Cooperation among the Five Eyes has provided the United States with significant intelligence benefits in all mission areas: counterterrorism, counter proliferation, cyber, regional challenges, and global coverage.

However, the longest intelligence organization on earth is the British Secret Intelligence Service (SIS) commonly known as MI6. Established in 1909 amid heightened and intensifying international rivalries when the British strategic policy makers were becoming especially concerned about the challenge of aggressive, ambitious, imperial

---

276     Acker, p. 335.

Germany, MI6 was tasked with collection and analysis of information about foreign foes and particularly Germany, which had shown its aggressiveness towards the British Empire with an objective to curtail its growth[277]. Indeed, by 1913 the British and French had formalized intelligence sharing and were exchanging intelligence materials about Germany[278]. Thus, the MI6 laid a foundation for intelligence sharing which continued till today.

The end of the Cold War was a remarkable period in the history of intelligence sharing among the world's nations. It witnessed a major upsurge in intelligence requirements and capabilities in response to a more challenging and more diffuse threat environment coupled with far-reaching technological advancements in information gathering and processing[279]. Further, the need for more intelligence exchange, increased aftermath of the 9/11 terrorist attacks on the United States, prompted a sharply increased appropriation of resources, federal and even provincial, for security and intelligence. Thus, the terror behaviour of the LRA brought international and regional action against it.

There are more theories which explain why intelligence sharing is vital. According to Liberalists, intelligence sharing improves decision making process. In the US the impact of intelligence sharing has been recorded within the Congress itself. It helps the congressmen to make sound decisions. Congress and other decision-making institutions need to share intelligence for a better and well-informed decision. This

---

277    Keith Jeffery. 2016. MI6: The History of the Secret Intelligence Service 1909-1949. Bloomsbury. P. 3.

278    Ibid, p.31.

279    Norman Hillmer & Maureen Appel Molot, eds., Canada Among Nations 2002: A Fading Power (Toronto: Oxford University Press, 2002, pp.141-171.

explains why the US, despite the end of the Cold War and the abundance of information available publicly or on a commercial basis, still has great need for intelligence and an in-house apparatus entrusted with its collection, production, and dissemination. Whatever the ultimate personality of the current phase of international relations turns out to be, it will not be an age of global peace and security. The past few years have witnessed classic aggression on a large scale as well as numerous instances of violence resulting from the breakdown of empires and states. Intelligence is essential if military personnel are to cope with such challenges, and it will continue to prove critical in helping government officials' fashion and implement policy in non-military realms that affect national security.

## Conclusion

In the fight against the LRA the importance and strength of intelligence sharing has never gone un noticed. This is not an isolated case. The end of the Cold war witnessed a major upsurge in intelligence requirements and capabilities in response to a more challenging and more diffuse threat environment coupled with far-reaching technological advancements in information gathering and processing Further, the need for more intelligence exchange, increased aftermath of the 9/11 terrorist attacks on the United States, prompted a sharply increased appropriation of resources, federal and even provincial, for security and intelligence.

# Chapter 8

<hr />

# Intelligence Sharing
# as a Pursuit of National Interest

## Introduction

This chapter brings the relationship between intelligence sharing, national security and national interest policy within in a general context. The Anti LRA Collison forces (UPDF, US Special force, the SAF and the SPLA) which opted for intelligence sharing as their first step forward for defeating LRA were all driven by their national interests. Museveni wanted stability of his government, South Sudan wanted stability but also wanted to pay back Museveni who stood with South Sudanese people during their difficult times and the US wanted to stop human rights abuses by the LRA.

## Overview of Intelligence Sharing and National Interest

Taking US experience, it is arguable that, neither the closeness imposed by globalization nor the need for intelligence sharing, is eliminated by new sources of information. Despite advancement in technology, there are still important dynamics that without intelligence sharing, are hard to learn facts about targets, including the intentions and capabilities of terrorists and criminal groups; unconventional weapons proliferation; efforts carried out secretly by unfriendly governments; and the disposition of hostile military forces. Such information, as Vice President Al Gore correctly states, is rarely available on the "information superhighway"[280] or through commercial satellite imagery; it is certainly not available with enough detail and timeliness to serve policymakers and combatants. Quite the opposite, there are a number of threats to US interests and well-being that can only be identified, monitored, and measured adequately by using dedicated intelligence sharing. This continuous search for information through greater paths dictates in some areas, a growing need for intelligence sharing that should not come as a surprise. The US government's creation of a modern intelligence capacity predated the Cold War. More than anything else, the desire to avoid another Pearl Harbour led to the creation of a centralized intelligence apparatus in 1947[281]. With this in mind, the US saw the need to avoid surprise attacks from hostile countries or groups still exists - for example Al Qaida of bin Laden, a Saudi businessman who turned terrorist.

---

280    The information superhighway was a popular term associated with former US Senator and Vice President Al Gore, and the term was popularly use throughout the 1990s to refer to digital communication system and internet telecommunications network.

281    See www.nationalww2museum.org.war

Moreover, the utility of intelligence collection and assessment relies on the continuing need to learn and get more from experience of others that are made available for the purpose of safety and security of individual and the entire humankind. To achieve this important task, intelligence collection is projected as the importance of sorting out mysteries, of analysing events and trends. This in turn, equips policy makers by increasing their understanding of the overall context, rather than in trying to predict individual events. Arguably, the cadre of intelligence analysts who are maintained and become part and parcel of the intelligence community, constitutes an important resource for policymakers trying to manage an enormous stream of information. By default, as much as by design, the intelligence community is increasingly becoming the locus for the US government, which relies on all sorts of information to help it design policies of physical and economic security.

The United States enjoys a position of unique power because of its military, economic, technology, diplomatic, and information might. As a result, it attained great influence in the post-Cold War world. This influence is extended to intelligence sharing, because intelligence itself is not simply the knowing, its sharing is also an important tool to enhance to security. One of the benefits enjoy by those nations which share intelligence is that it enables them - be they friendly governments, alliances, or institutions like the International Atomic Energy Agency and other UN agencies - to be more effective in dealing with common challenges. From the United States perspective many multilateral efforts will succeed only if the United States possesses and is willing to share the necessary means. As Larry D. Thompson explains:

> *The collection of intelligence within the United States is obviously a critical component of the federal government's efforts in the war against terrorism... The first step in disrupting terrorist plots and preventing terrorist attacks is obtaining information about the operations of terrorist groups and the activities of individual terrorists, both inside and outside of our nation's borders.*[282]

Truly, intelligence collection and sharing is only the first step in combating terrorism, and a piece of information is like a piece of a puzzle[283]. Arguably, it is only when a piece of information is combined with many other pieces of information does the big picture - for instance defeating terrorists - emerge. Moreover, it is only the utilization of information rather than possession of information that can stop terrorism; in this sense, it is argued that information must lead to action which disadvantages the terrorists. This is why it is critical for the US, Uganda, and South Sudan to share intelligence on LRA activities in the same way information is shared in the US with other components of the federal government who possess similar information and leads to preventive action which protects American lives. Inability to share intelligence brings serious consequences for example before the horrific terrorist attacks of September 11, 2001. As Thompson argued, he witnessed a grave deficiency among intelligence officials and law enforcement officials to share information with each other, which hampered the Department of Justice's ability to act to defend the nation

282    Thompson, L. D. "Intelligence Collection and Information Sharing within the United States" Monday, December 8, 2003, https://www.brookings.edu/testimonies/intelligence-collection-and-information-sharing-within-the-united-states/
283    Ibid.

against terrorist attacks.[284]

While the US was able to bring Uganda, Sudan and South Sudan into intelligence sharing agreement on the LRA, it did not prevent contradictions and challenges brought by this arrangement which existed among these countries. These countries were not unified by one interest, therefore, their intelligence priorities were not the same, since each had vital national interests as dictated by the priorities of their own national security policy. Interests matter most to a country. Thus, policy priorities, while taking the inherent importance of interests into account, must also reflect existing anticipated threats as well as opportunities in various areas of politics, economics, and defence. Furthermore, modern systems for intelligence collection are expensive, and the demands on the intelligence community from policymakers and the military to collect and assess information for a wide array of tasks, have been growing. Accurate intelligence significantly improves the effectiveness of diplomatic and military undertakings; while good intelligence cannot guarantee good policy, poor intelligence frequently contributes to policy failure. States will have to continue to devote significant resources to this area if it wants an enhanced capability and population protection at large.

Indeed, historical events have an undeniable influence on geopolitical relations of Sudan, South Sudan and Uganda. Thus, it is impossible to understand the close ties that existed between South Sudan and Uganda and likewise, it is impossible to understand close ties that existed between Sudan and the LRA without being aware of geopolitics of these countries. The geopolitics has always been dominated by issues of conflict and cooperation relationships between and among these countries.

---

284    Ibid.

The immediate significance of intelligence sharing has been the reduction of LRA capacity following UPDF and SPLA incursion, as directed by the US intelligence officers into the LRA hideouts in South Sudan and DRC. These attacks forced members of the LRA, including senior ranking officers, to abandon rebellion and a few others among them were killed. More importantly, and in relation to these campaign attacks supported by intelligence sharing arrangements, civilian killings by the LRA also decreased markedly in the years following, for example from 1,200 in 2009 to fewer than 20 in 2014.[285] The number of people displaced as a direct result of LRA attacks, or out of fear of coming under attack by the LRA, has also declined. As of mid-2015, almost 200,000 people were estimated to be internally displaced or living as refugees in CAR, DRC, and South Sudan due to LRA activity, compared to over 326,000 reportedly displaced as of December 2013. By 2014, the LRA manpower declined to a small, dispersed armed group active in remote areas of Central Africa.

Thanks to US intervention, the LRA 's infliction of widespread human suffering and its potential threat to regional stability have drawn significant attention in recent years, including in Congress. Campaigns by US-based advocacy groups have contributed to policy makers' interest. Since 2008, the United States has provided support to Ugandan-led military operations to capture or kill LRA commanders, which since 2012 have been integrated into an African Union (AU) Regional Task Force against the LRA. The Obama Administration expanded US support for these operations in 2011 by deploying US military advisors to the field. In March 2014, the Administration notified Congress of the deployment of U.S. military aircraft and more personnel to

---

285    Baguma. P. 12

provide episodic enhanced air mobility support to African forces. The United States has also provided humanitarian aid, pursued regional diplomacy, helped to fund early-warning systems, and supported multilateral programs to demobilize and reintegrate ex-LRA combatants. The Administration has referred to these efforts as part of its broader commitment to preventing and mitigating mass atrocities. Growing US involvement may also be viewed in the context of Uganda 's role as a key US security partner in East and Central Africa. US security assistance to Uganda, including for counter-LRA efforts, has continued despite policy makers' criticism.

Although the pursuit of national interests may generate conflict depending on the nature of cooperation, nevertheless, cooperation remained the benchmark foundation of intelligence sharing. Isabel et al developed a model of three basic forms of interaction in conflicts situation. These forms are the cooperative interaction, conflictual interaction, and dominating interaction[286]. Collins, further suggested that the interaction consists of ingredients[287]. First, there must be co-presence of at least two persons in the same place so that they affect each other by their bodily presence, whether or not they are aware of it. Second, there must be barriers to outsiders where participants have a sense of who is or not taking part in the interaction. Third, the people must focus their attention on the same activity or object and become aware of each other's focus and fourth, people must share a common mood or emotional experience. Indeed, we live in a world in which

---

286     Isabel Bramsen, Poul Poder and Ole Waever (2019). Resolving International Conflict: Dynamic of Escalation, Continuation and Transformation. Routledge. p.38.

287     Collins, Randall (2005). Interaction Ritual Chains. Princeton University Press. p.41.

social and political conflict emerge suddenly and unexpectedly[288].

The Osama bin Laden's organization of a surprise attack on the United States twin towers was indeed an unexpected, sudden and disastrous. Thousands of people died as a result of this act of terror. Subsequently, it was followed by wide condemnation never ever seen before, with US and other major developed countries calling for a collaborative War on Terror in general and on al-Qaida specifically. But more significantly, it rejuvenated intelligence sharing to the centre stage of international community. As Elizabeth Sepper correctly argues:

> *"Although the global community—and the United States in particular—had long discussed improving information sharing to combat transnational crime, the events of September 11, 2001 brought networks for information sharing centre-stage"[289].*

In relation to this, it is understood that the 9/11 events were the turning point in the history of intelligence sharing because of its greater focus on intelligence sharing. With Sudan supplying the US with information on suspected terrorists, it did not take long for Sudan to handover about forty members of al Qaida suspected of life attempts on President Hosni Mubarak to the Egyptian government[290]. US government had to rely on Sudan because the Sudan ruling NIF party was implicated in the assassination attempt on the Egyptian President

---

288    Isabel Bramsen, Poul Poder and Ole Waever (2019). Resolving International Conflict: Dynamic of Escalation, Continuation and Transformation. Routledge. p.38.

289    Elizabeth Sepper. Democracy, Human Rights, and Intelligence Sharing TEXAS INTERNATIONAL LAW JOURNAL [VOL. 46:151

290    Hilde, p.19.

Hosni Mubarak in 1995. Also, Bin Laden - the al-Qaida leader, had been based and had his militants training schools in Sudan. All these made the US government believe that Sudan was a friend to terrorists.

## Contribution of 9/11 to Intelligence Sharing

Although 9/11 was a sad incident which cannot be forgotten in human history, it nevertheless contributed to advancement of intelligence sharing. This time, there was a rallying of countries against terrorism. On that day of 11 September 2001, nearly three thousand Americans and non-Americans lost their lives during this terror attack. And over twenty-five thousand innocent people were also wounded. Immediately, the incident was followed by condemnation all over the world. There was also declaration of a 'War on Terror', spearheaded by the United States. Thus, 9/11 was very significant in the sense that it united and brought awareness among world leaders of the danger of terror, as terror has no borders. In 1998, terror groups demolished the US Nairobi Embassy and attacked the US Embassy in Dar es Salam causing a significant amount of death and injuries. A number of terrorists' attacks were also registered in other parts of the world. This range of coverage by terrorists' activities rang an alarm bell that something more serious was on its way.

9/11 contributed to the realization that the only way to address and control expansion of terrorism was for the world to adopt a unity of purpose. This unity would be directed against terrorist organizations. It is only through this unity of purpose that human security is possible. Through collective efforts such as by sharing information on the movements of terror groups, nations would be able to predict and to prevent further deadly strikes such as the one on twin towers in the

US by Al-Qaida. In the aftermath of 9/11, nations were now able to realize the weakness of non-cooperation and its adverse impact on human life.

Therefore, it is imperative to say, unless there is intelligence sharing and exchange of information between and among the nations that would horrific manmade disasters such as the killing on US twin towers, the US Nairobi Embassy and Dar es Salam Embassy be prevented. Unless nations cooperate through intelligence sharing, would it be possible to defeat terrorists, especially when they operate on irregular basis? The rationale behind a terrorist's irregularity is signified by their lack of the military capacity to defeat their opponents on the battlefields. To compensate for this weakness, they exploit the military advantages of launching surprise and terror attacks on major economic and military free installations, mostly concealing themselves in unexpected terrain such as among the civilian population[291]. They operate like guerrilla forces, the only different between them is that the arena of guerrilla forces and operations is internal, while terrorist organizations operate at the international level. It is also correct that some terrorists operate internally. This wider range of operations has left governments with only one choice. Governments can effectively encounter terror forces by use of surprise and concealment only if they have accurate intelligence on their opponents' plans and bases.[292]

Accurate intelligence is simply understood to mean the collection, protection and analysis of publicly and secret available information for a purpose of disrupting terror activities to avoid harm. Whatever

---

291    James Igoe Walsh. Intelligence Sharing for Counter-Insurgency. Defense and Security Analysis Vol. 24, No. 3, pp281-301, September 2008.

292    Ibid.

comes out of states' co-operation in the process of collection of intelligence and sharing must translate itself into a practical benefit of a better understanding of the strengths and weakness of the various strategies for countering insurgencies[293]. The transnational nature of several terrorist organizations, with al-Qaeda being the most notorious, implies that their detection, disruption, and elimination can succeed fully and only if done through globally and internationally coordinated effort in the form of intelligence sharing.

## Conclusion

Intelligence sharing at the wider context is the only grantee of national safety and international. The transnational nature and the wide range coverage of several terrorist organizations, implies that their detection, disruption, and elimination can succeed fully and only if done through globally and internationally coordinated effort in the form of intelligence.

---

293     Ibid.

# Chapter 9

◇◇◇◇◇◇◇◇◇◇◇◇◇◇◇◇◇◇◇◇◇

# The International
# Intelligence Sharing Before 9/11

## Introduction

It is not suggested that international intelligence cooperation had never existed prior to 11 September, rather it is only argued that the 9 11 changed the perception how the intelligence was handled. In fact, Western security and intelligence agencies have long cooperated, either bilaterally or multilaterally. However, their cooperation is sometimes difficult, uneven, and haphazard, but when lives are believed to be at stake due to terrorists' active targeting, efforts to make it work are certainly redoubled[294]. Indeed, in the aftermath of 9/11 the US saw not

---

294     Stéphane Lefebvre. The Difficulties and Dilemmas of International Intelligence Cooperation. International Journal of Intelligence and Counterintelligence, 16: 527–542, 2003.

only its NATO counterparts rise to action, but also a new enthusiasm from its traditional bilateral relationships in improving counterterrorism coordination and more specifically, intelligence sharing[295]. On this collective effort, Derek Reveron states "the war on terror requires high levels of intelligence to identify a threat relative to the amount of force required to neutralize it"[296]. Janine McGruddy also suggested that "since 9/11 the range of partners in the intelligence world that share information at the international level has grown exponentially."[297]

## International Networking of Terrorism

Economic crisis and other unfavorable conditions in Africa, the Arab world and Asia are viewed to increase networks of terrorism. Terrorism is a political international movement whose membership consists of desperate groups brought together by pressures of poverty, and who are looking for ways of survival. Thus, terrorists derive their membership and recruit from these desperate groups, especially from those who left their homes in Africa, Middle East and Asia, Latin America to the developed world in search of better life, which turned out to be frustrating when Europe and America turn to be opposite of their preconceived expectations. Having already dealt away with their little households in their home countries to facilitate their movements to their expected green pastures through smugglers, these groups

---

295    Anna-Katherine Staser McGill and David H. Gray Challenges to International Counterterrorism Intelligence Sharing Global Security Studies, Summer 2012, Volume 3, Issue 3, p.77.

296    Reveron, D.S. (2006). "Old allies, new friends: intelligence-sharing in the war on terror." Orbis 50 (3): 453–68.

297    Janine McGruddy Multilateral Intelligence Collaboration and International Oversight Journal of Strategic Security Volume 6, Number 5 Volume 6, No. 3, Fall 2013, p. 214.

reached a point of no return, their only one choice is to go to Europe or America how much risky it cost.

Occasionally, such illegal groups seeking new lives in Europe or US are most the time intercepted by the border guards of those transit nations with help from those countries targeted by the migrants. This is intended to prevent entrance of those desperate migrants who are looking better opportunities of life outside their own home areas. Unfortunately, the transit nations tend to be caught unprepared by the migrations, hence, face the burden of accommodating the illegal migrants, until when they take the decision to decide on what to do with them. While undergoing this process of granting them refugee status or not the migrants experience harsh and difficult conditions during these temporary camps which may take a very long time, until they are finally either deported to their places of origin or granted refugee status. Returning to their countries of oration disappointed, some members of these groups opt to join any radical groups which they come across.

All the illegal migrations discussed above so far share a common feature, in that they were often facilitated or at least accompanied by illegal financial transactions that take place through formal or informal channels handled by groups of human trafficking mafia. During the conflict, the LTTE collected vast sums of money in various countries around the world and transferred these funds to Sri Lanka through various legal and illegal channels. Tracking these transfers, particularly when they happened through informal channels, was a very difficult exercise. Having a proper understanding between the Financial Intelligence Units, Intelligence Agencies, Law Enforcement Agencies within the region is therefore extremely important in tracking illegal financial transactions and in identifying and apprehending the culprits involved. Particularly in today's globalized context, the full potential of regional organizations such as to enhance technical and security cooperation must be exploited by member nations.

Although there are international agencies that are engaged in these activities, experiences have shown that bilateral and regional cooperation amongst the intelligence and law enforcement institutions of nations have produced effective results. Ultimately it is the cooperation, mutual assistance and cordial relationships within the region that will help nations in the region to achieve their goals. The LRA atrocities directed against civil population in northern Uganda, South Sudan, Central Africa and Democratic Republic of Congo (DRC) appeared as a regional threat - with Uganda, South Sudan and the Democratic Republic of Congo (DRC) championing joint operations against the LRA through a trilateral arrangement in December 2008[298]. However, logistical and political challenges prompted the coalition to request AU support for a strong Regional Task Force (RTF) to pursue LRA. This prompted the US to extend its hand, eventually leading the United States to spend almost $800 million on the effort since 2011, when President Barack Obama deployed Special Operations forces to the region to provide advisory support, intelligence and logistical assistance to African Union soldiers fighting the Lord's Resistance Army.

In the last couple of decades, states have demonstrated an increased willingness to participate in threat intelligence sharing platforms. This required cooperation in the use and flow of sensitive information as

---

298     United Nations Security Council (UNSC). 27.3. 2009. 'Twenty-seventh report of the Secretary- General on the United Nations Organization Mission in the Democratic Republic of the Congo'. S/2009/160.

well promoting awareness among governments. It further required maintaining clarity in agreements and endorsing jurisdictional limits to ensure international cooperation, in order to bring out the best benefits for protecting our world.

This increased open exchange of information and knowledge regarding threats, vulnerabilities, incidents and mitigation strategies results from the states' growing need to protect themselves against today's sophisticated terror attacks. This essence of intelligence sharing in the War on Terror has contributed to policy development especially in the US. The Bush administration's designation of its national strategy as a 'War on Terror' highlights the importance of combating terrorism on an international level[299]. Fundamental to this effort is bilateral intelligence sharing. Intelligence reform efforts to date have focused on improving intelligence sharing within the US intelligence community. However, critical intelligence can be gained through America's international partners.

Officials from the countries of US, Uganda and South Sudan involved in this arrangement acknowledged that they have significantly degraded the LRA, diminishing it to around 100 people today from a fighting force of 3,000[300]. In an interview with the New York Times General Harrington of the US Army remarked that:

> It's time to go home... the mission to get Mr. Kony could be looked
> at, in retrospect, as a mission "to remove a regional threat," ...the

299     Reveron, D.S., 2006. Old allies, new friends: intelligence-sharing in the war on terror. Orbis, 50(3), pp.453-468.

300     Hellene Cooper interview with Maj. Gen. Joseph P. Harrington, the commander of United States Army Africa
May 15, 2017.

*operation has been successful, degrading the Lord's Resistance Army to
where it is now. ... "LRA is really no longer a relevant organization.*[301]

Before intelligence sharing took effects in 2002, the LRA had been a major threat which extended beyond Uganda's boundaries. Estimates by human rights groups calculate that at the height of the violence, more than two million people were forcibly relocated to internally displaced persons camps, tens of thousands of civilians including men, women, and children had been abducted and uncounted thousands more killed by the LRA and many villagers suffered from persistent LRA attacks[302]. Similarly, The Clingendael Institute in its 2008 reports, suggested that northern Uganda had unquestionably suffered the greatest burden of LRA activity. By 1987, tens of thousands of people had been killed and 1.8 million were displaced by the LRA war. They also noted that the LRA had gradually become a threat to other countries in the region[303]. The same report suggests that the rebel group has been able to move freely, threatening security and stability in southern Sudan, DRC and the Central African Republic (CAR)[304]. Similarly, the UN Office for the Coordination of Humanitarian Affairs (OCHA) stated that in 2009 the LRA had killed 1,096 civilians and abducted 1,373 adults and 255 children in the Haut and Bas Uele districts of northern Congo alone.[305]

---

301     Ibid.

302     Northern Uganda Transitional Justice Working Group.

303     Hemmer, J. (2008). The Lord's Resistance Army: tackling a regional spoiler. Clingendael Institute, p. 1.

304     Ibid.

305     Human Rights Watch (2010) Trail of Death: LRA Atrocities in Northeastern Congo. New York: Human Rights Watch, p. 17.

However, following intelligence sharing, the LRA activities were reduced to insignificant threats. As Sylvester *remarks:*

> *In December 2008, in a feat of reprisal for the failed FPA, the Uganda army, (UPDF), DRC Forces, Armees de la Republique Democratique du Congo/Armed Forces of Democratic Republic of Congo (FARDC) and GoSS (Sudan People's Liberation Army or SPLA) launched a joint military offensive, Operation Lightning Thunder, targeted largely at north eastern DRC in the Garamba Forest, where the LRA had set up base. The offensive weakened the LRA, cut off supplies, and scattered the LRA across the region[306]". Another effect of intelligence sharing on the LRA - and as reported by the LRA crisis tracker, has been the reduction in its fighting strength. The Resolve states "in recent years, as its fighting strength has diminished, LRA groups have evolved to avoid detection by troops from the AU RTF, US military, and UN peacekeeping missions. Large LRA groups have fragmented, and LRA commanders have increasingly sought to combine traditional looting attacks with less violent and therefore less conspicuous survival strategies.[307]*

As the situation of the LRA has shown, intelligence sharing has been an effective tool to save life of people from suffering. Its importance first increased during the two World Wars and intensified during the Cold War. Shared interests in intelligence sharing during the Cold

---

306     Maphosa ,S. B.. The Lord's Resistance Army: A Review of African Union Regional Efforts to Eliminate the Resistance in Central Africa in Festus B Aboagye Ed. The Comprehensive Review of the African Conflicts and Regional Interventions, 2016. P.196.

307     The Resolve LRA Crisis Initiative. (2016). The State of the LRA in 2016, p.8.

War influenced the next period in the evolution of US foreign intelligence partnerships. Similarly, during the Cold War, intelligence sharing in NATO was strategic in nature and focused intensively on political and military factors, including sporadically, the economic sector. Following the world's concern from the effects of the 9/11 terror attacks in the United States, intelligence cooperation became even more relevant and a necessity with decision makers, who were now required to make intelligence sharing an essential element in the global fight against international terrorism.

In 2002, Uganda, and Sudan signed a protocol allowing Uganda Forces to enter Sudan (now South Sudan) territory to hunt down the LRA, which had been fighting President Yoweri Museveni government since 1987[308]. The LRA had bases in South Sudan from where they launched their operations into Uganda, and had operated as friendly troops to the Sudan Armed Forces (SAF). By 2002, when the protocol for hunting down the LRA was reached by Uganda, Sudan and US the Comprehensive Peace Agreement (CPA) that granted the Government of southern Sudan (GOSS) was still being negotiated and had not been concluded. Thus, the Sudan government was the main signatory to the protocol. The protocol required the armies of Uganda - Uganda People's Defence Forces (UPDF) and Sudan Armed Forces (SAF) to share intelligence on LRA. SPLA was not part of this arrangement although it collaborated with UPDF as a matter of fact.

The protocol that led to intelligence sharing between Uganda and Sudan was broken by the US. Before the US could bring Sudan and

---

308    Agreement between the Governments of Sudan and Uganda, December 8, 1999 http://www.cartercenter.org/documents/nondatabase/nairobi%20agreement%20 1999.html

Uganda together to exchange intelligence against the LRA, tensions existed between Uganda and Sudan over accusations and counter accusations over support of dissidents. Sudan accused Uganda of supporting the SPLM/A to overthrow its government in Khartoum and Uganda in turn accused Sudan of providing military logistics to the LRA to overthrow its government in Kampala. Neither Sudan nor Uganda was wrong. Both countries supported dissidents.

### US, Uganda, South Sudan and Sudan Intelligence Cooperation

This new deal of intelligence sharing relations was seen as part of the United States diplomatic efforts to reconcile Uganda and Sudan, so that the two were able to finish off the LRA. This was indeed a remarkable move in the history of Sudan and Uganda relations, unfortunately, it put Sudan in a dilemma: either cooperate, or continue supporting the LRA now black listed by the US. A rationalist Sudan accepted to cooperate, to avoid action from the US. Obviously, this was the best decision to make. This move by Sudan was intended to please US, whose eyes were on Sudan because of its connection with terror organizations.

The late comer to intelligence sharing agreements on the LRA was the Government of Southern Sudan. Following the conclusion of the CPA which resolved Sudan's internal conflict between SPLA in the south and the government in the north in 2005, the Government of South Sudan (GoSS) based in Juba, decided to join intelligence sharing agreement against the LRA. South Sudan became very significant in terms of human intelligence because of its knowledge of terrain and because the LRA operated in southern Sudanese territory. The US forces relied on signal and communication intelligence but the remote nature of southern Sudan, combined together with its equatorial thick forest terrain, made the need for human intelligence paramount. The human intelligence provided by South Sudanese forces, and the US communication intelligence reinforced by the UPDF combat operations, undeniably cleared LRA activities from Uganda and South Sudan and pushed the LRA to the bush of DRC and CAR, where its remnants are believed to be hiding to this date.

Before intelligence agreement and the involvement of US Special forces and the SPLA South Sudan; Uganda has always pursued military options sometimes with disappointment. The LRA reacted violently to the UPDF operations by striking heavily on civil population targets in the direct sight of the UPDF generals and their soldiers, making UPDF moves meaningless. UPDF has always justified its action on the basis of security of person and their assets. However, security of persons and their assets is better protected under the intelligence sharing agreement.

## The US Factor in the Intelligence Sharing Agreement

The significance of the US in the trilateral intelligence sharing agreement between Uganda, South Sudan and Sudan is seen in its ability to force Sudan to join the LRA war. Through US engagement, Sudan was able to accept sharing intelligence with Uganda on the LRA, something that has been difficult before the US involvement. The US involvement led to intelligence sharing. Intelligence sharing was mainly to find the location of Joseph Kony in particular and stopping Sudan from providing logistical support to the LRA. The sharing of intelligence information involved different agencies within the government of Uganda, South Sudan, Sudan and the US, noting that "the ability to exchange intelligence, information, data, or knowledge among them as appropriate as possible will bring the LRA war to an end"[309].

Indeed, the coming together of the US, Uganda, South Sudan and Sudan to fight LRA through intelligence sharing proved to be a good thing. Indeed, through intelligence sharing, Sudan was able to relinquish its ties with LRA, which served the Sudan armed forces fighting the SPLA in the south. It is recalled that these two neighbouring countries have had deep-rooted historical conflicts throughout their period of independence. This conflict has been due to accusations and counter accusations about support to rebels as a one way of achieving disagreements over border issues.

While the US has succeeded rhetorically to bring the two countries together in sharing intelligence and fighting the LRA, it is equally argued that the US had underestimated the degree of mistrust between Sudan and Uganda, and the two continued to be closed throughout the process of their intelligence sharing and instead, they turned the whole process into a war game. According to Major Shaban Bantariza, the UPDF spokesman, the Ugandan military had finished the LRA by 1992. However, in 1994 Sudan revived the defunct rebel movement by providing it with arms and allowing the LRA to establish bases in southern Sudan[310]. Van Acker on his part, suggests that there are three possible groups of answers to questions concerning the survivability of LRA, and all containing an element of truth[311]. His suppositions

---

309     The Importance of Intelligence Sharing among Nations. Asian Tribune. 8 May 2013.

310     See Titeca, K. and Costeur, T., 2015. An LRA for everyone: How different actors frame the Lord's Resistance Army. African Affairs, 114(454), pp.92-114.

311     See Van Acker, F., 2004. Uganda and the Lord's Resistance Army: The new order no one ordered. African Affairs, 103(412), pp.335-357.

are that either the LRA is strong, because it is supported by the local population and/or supported from the outside. Or the UPDF is weak, because of internal failings and/or competing demands placed upon it. Or alternatively, the war in the north is a façade for other goings-on, such as a wish to support the SPLA at a political level, and/or doing well out of war at the level of individual Ugandan army officers and commanders, just as the war provides opportunities for politicians to posture in the role of peacemakers[312].

Furthermore, the coming in of the US at least psychologically shaped the thinking in the region. Indeed, when the intelligence sharing agreement was signed nearly 8 years ago, it was thought that the LRA would be wiped out in a few months. Indeed, progress was made against the LRA by this arrangement, however, the LRA continues to exist and carries out looting missions mainly targeting civilians in DRC and CAR. This point has been emphasized by Atkinson, who thinks that dialogue should be pursued in order to solve the LRA problem. Atkinson states:

> *Pursuing a military solution to the LRA problem has failed for two decades and is unlikely to be successful now...the only feasible is to approach attempt to re-establish peaceful dialogue ...[313]*

Re-establishing peaceful dialogue would mean examining reasons that hindered Juba peace process. It was observed that the peace talks did not get off to a promising start because of reluctance from the international community, and the parties' mistrust of the chief mediator

---

312    Ibid.

313    Atkinson. p.1.

Dr. Riek Machar[314]. Issues of mediator trust and international community commitment have proved to be central concepts in the area of peace agreements and conflict resolution. The balancing between peace and atrocities committed by the LRA must also be considered whenever we think of ending the LRA conflict.

There has been no reasonableness in the LRA war when looking at the relationship with civilians. Indeed, one of its evils was the increase in human trafficking across the borders of Uganda, South Sudan, CAR and DRC. Primarily due to LRA brutality, large numbers of people from these countries were abducted and forcefully converted into LRA forces, with some young girls unwillingly becoming wives of LRA senior members including Joseph Kony himself. Due to these social reasons, abductions became a lucrative business for LRA fighters.

According to Jennifer E. Sims, since the terrorist attacks of 11 September 2001 (9/11), the role of foreign intelligence sharing took centre stage in the global War on Terror[315]. He further argued that intelligence acquired through foreign liaison is widely credited with having helped thwart attacks in Bahrain, arrest al-Qaida leaders in Pakistan, and chase complicit Taliban from hideouts in Afghanistan[316]. Indeed, the obvious impact of intelligence sharing by the US, Uganda and Sudan, and South Sudan in the fight against the LRA has been their ability to push away and reduce the LRA activities within the borders of Uganda. This in turn, led to the reduction of LRA abuses on the civil population. However, it is yet to be established how

---

314    See Mareike, p. 34.

315    Jennifer E. Sims. (2006). Foreign Intelligence Liaison: Devils, Deals, and details. International Journal of Intelligence and Counterintelligence. Routledge. Vol 19, Pp. 195-217.

316    Ibid.

intelligence was acquired, generated and used. Much contemporary study of intelligence is concerned with how knowledge is acquired, generated and used[317]. We know that intelligence is information not publicly available, or analysis based at least in part on such information, which has been prepared for policymakers or other actors inside the system. Intelligence is unique for its use of information, in that it is collected secretly and prepared in a timely manner to meet specific needs, such as eliminating LRA and its leader Joseph Kony.

While it may be perceived a blessing to bring unequal partners such as the United States, Uganda, Sudan, and South Sudan into collaboration, it is equally challenging to sustain the collaboration once the stronger denominator quits, as this very case as shown. In addition, some of the members in this arrangement, particularly Sudan, joined to please the US and improve its sour relations with the US. Therefore, its participation was never genuine. On its part, Uganda was very reserved about Sudan, since it viewed Sudan to be the godfather of the LRA and other pre LRA rebel groups in Uganda. As President Museveni states: "the insecurity in Uganda was caused by terrorists supported by the Sudan"[318]

In the LRA intelligence sharing, the US was the denominator that progressed the operation financially and technologically. In October 2010, US Government sent 100 special forces to help Ugandan troops in logistics and also to provide military advice on the LRA. After the formation of AU RTF, the US decided to become a partner and added in additional 50 military men, making the total US forces involved

---

317    Len Scott. (2006). Secret Intelligence, Covert Action and Clandestine Diplomacy, Intelligence and National security, 19:2, 322-341.

318    Museveni, p. 256.

in the fight against the LRA 150 persons. These troops brought with them intelligence and communication equipment which made intelligence and information gathering and sharing possible.

The irresponsible behaviour the LRA had demonstrated in its fight against Museveni and his government manifested itself in serious human rights violations mostly in northern Uganda and South Sudan. This caused many internally, regionally, continently and internationally to see that the LRA was a movement without a justifiable cause; and therefore perceived as an international threat to peace and security. Consequently, it prompted an international move to address LRA behaviour. Various options and moves were considered against the LRA, including intelligence sharing among states affected by the LRA insurgency, and in order for or this move to be effective, it necessitated developing the intelligence needed to fight the LRA effectively.

## The Obama Administration's Policy Towards the LRA

In 2010, President Barak Obama pushed and signed into law, the LRA Disarmament and Northern Uganda Recovery Act of 2009, which was essentially designed to be the policy endeavour to support the stabilization and lasting peace in northern Uganda and areas affected by the LRA, through development of a regional strategy to support efforts to successfully protect civilians and eliminate the threat posed by the LRA[319]. It also aimed to authorize funds for humanitarian relief and reconstruction, reconciliation, and transitional justice, and for other purposes. Accordingly, in October 2011, the US government

---

319    Sylvester B. Maphosa. The Lord's Resistance Army: A Review of African Union Regional Efforts to Eliminate the Resistance in Central Africa, in A Comprehensive Review of African Conflicts and Regional Interventions 2016. P. 212-264.

deployed a team of around 100 military experts to Uganda, South Sudan, CAR and the DRC to assist UPDF, SPLA and other regional forces in defeating the LRA insurgency. The policy also aims to adopt other means to defeat LRA; including encouraging defections from the LRA, and protecting civilians.[320]

Secondly, the fact that the actual capacities of the LRA were unclear, even for those within the LRA, increased the space for a wide variety of interpretations. The malleable identity of the LRA combined with external pressure and foreign aid made the movement an issue which served a variety of functions and interests. In the words of a Kampala-based donor actor, the LRA became "a political football to kick around with rather than an issue as such".

Because intelligence work is a product of team effort, there are certain peculiarities common to the bureaucratic environment, such as in the case of intelligence sharing on the LRA, which help explain either success or failure. There are those who argue that the government's failure to capture and defeat the LRA was intentional[321]. They explain that the Museveni government was unwilling to end the war because it served the administration's own interests and that there were political reasons for allowing the conflict to continue. There are those who feel that Museveni government has been using the LRA war to prevent political mobilization that could bring about the end of the president's reign. Many individuals, especially from the north, believe that the war was a strategy to camouflage a slow genocide aimed at eliminating them as a people. However, such feelings may not hold the

---

320     Ibid.

321     Ahere, J and Maina Grace. The never-ending pursuit of the Lord's Resistance Army: An analysis of the Regional Cooperative Initiative for the Elimination of the LRA. ACCORD, Issue 024, March 2013.

ground since Uganda forces were able to demonstrate their ability to push LRA away from their strong holds in northern Uganda into the bush of CAR and DRC.

However, this support would not be possible without support provided by the US. The US government provided intelligence experts to Uganda forces. With President Museveni's government now receiving military aid and diplomatic support obviously it was in strong position to defeat the LRA. However, this support was never without bill to pay. It was not a free lunch. In exchange, Museveni served as a conduit to the Sudan People's Liberation Army (SPLA) which had been at the forefront of waging war on the Khartoum government. Recently, the US sent 100 Special Forces to hunt for Kony. Many Ugandans are opposed to the involvement of the USA as they believe that it will result in the death of more innocent civilians. The Museveni regime continues to spend huge amounts of its budget on the military, justifying this expenditure as being essential to the war effort against the LRA. There have been claims that the LRA had some sort of support from the Acholi population, both locally and abroad. These claims, however, have been largely unproven and it is doubtful that if the LRA was acting as an agent of the local community, they would inflict suffering on the very people they were acting for.

The LRA is regarded as a well-trained and equipped armed group with a strong command structure and perseverance. While the LRA cannot boast of its dominance and strength today, the group's ability to adapt to the harsh terrain in which it operates indicates that a more consolidated and well processed plan of action is needed if the group is to be defeated. The movement's ability to adapt is evidenced by its survival and expansion into the DRC and CAR. In previous engagements the UPDF was, on the whole, less prepared in comparison to the LRA. It was the UPDF's lack of capacity and training, poor morale, competing engagements in the DRC and grand-scale corruption that significantly contributed to the success of the LRA. The war has become a lucrative source of income and wealth for certain key individuals. High-ranking military officers, government officials and powerful LRA rebels benefited as a result of the war.

This raised questions regarding whether both the government and the LRA were committed to ensuring an end to the war. A good illustration of how the government profited, is evidenced by information in the army pay rolls which showed payment to about 10,000 'ghost' soldiers. The Ugandan government has been a regular recipient of donor support, which makes up half of the national budget. Some of the funding received is meant for spending on defence, with the primary intention being to strengthen the UPDF's capacity to exterminate the LRA. Donor support is mainly received from bilateral development partners, especially under the ambit of the Northern Uganda Reconstruction Program who have been focusing on achieving an end to the long-drawn conflict in northern Uganda. Given the alluded corruption involved in administering donor funds in Uganda, a question may be raised regarding whether the donor funding is contributing to improving the situation or perpetuating the conflict. The donor community and the funds they disburse present a questionable dynamic for those working to end this conflict. Receipt of some donor funds could in many ways have contributed to inhibiting efforts to end the conflict.

## Challenges in Intelligence Sharing on the LRA

As sharing intelligence agreement proceeds with implementation, contradictions between members of the agreement emerged. These contradictions in the intelligence sharing arrangement were mainly encountered in intelligence collection priorities, more importantly between Sudan and Uganda simply because these very two members in intelligence sharing have always been in conflict over support of oppositions. As Frank obviously reiterates: "common wisdom used to be that the reason for the LRA's survival was Sudanese sanctuary".[322]

These contradictions were latter on to become a factor in the search of Joseph Kony and for this reason Joseph Kony remains at large till today. The inability of the parties to the intelligence sharing agreement to capture Joseph Kony is thus, an intelligence failure. An intelligence failure can be defined as any misunderstanding of a situation that leads a government or its military forces to take actions that are inappropriate and counterproductive to its own interests[323]. The failure to attain intelligence accurately that would have led to capturing of Joseph Kony and compel the rest of LRA members to accept peace as initially foreseen, can be explained in terms of an intelligence error of underestimation or overestimation.

While intelligence may be collected with concentration and diligence, nevertheless, it is a big mistake for anyone to think that any human undertaking, including collection of intelligence, will be error-free. Enemies may be underestimated or overestimated, and events that should be predictable go unforeseen. However, this has not

322    See Frank Van Acker (2004). Uganda and the Lord's Resistance Army: The New Order No One Ordered. African Affairs. Vol. 103. No. 412, July, pp.335-357.

323    Schulsky, A. & Schmitt, G. (2002). Silent Warfare: Understanding the World of Intelligence. Washington DC: Brassey's., p.63.

been the case with intelligence sharing against the LRA. It must be remembered that Sudan in the first place has never been willing to part completely with the LRA given the LRA's position of proxy.

The nature of the operations on the LRA by the US, Uganda, and South Sudan was a clandestine one, of which the most important method of intelligence collection to be relied on was the human intelligence, even though such intelligence was complemented by other sources such as signal and imaginary, which the involvement of the US made possible. Unexpectedly, this was not the case in this situation. The US relied on its advanced technology and ignored the important factor of human intelligence, which the SPLA was good at, considering its knowledge of terrain combined with the factor that its members were among the same locals as the LRA. In this way, the chance to trace Joseph Kony's location was missed. Human intelligence can also help shed light on intentions as well as capabilities, in this case on the capabilities and the intentions of the LRA senior line. Such knowledge was likely to prove crucial in tracking the activities of the LRA and in determining the status of their future activities, unfortunately little attention was paid to local knowledge.

Throughout their activities towards the LRA operations, the US, Uganda, Sudan and South Sudan were an unable to take actions that were appropriate and productive to their own interests of killing or capturing Joseph Kony[324]. What was required to meet end needs of getting rid of Joseph and his close generals, was the full and unwavering commitment from Sudan and use of local knowledge by collaborating with indigenous people through the SPLA. Because SPLA was the weakest partner, little attention was given to it in any aspects - be that finance, training or equipment needed for intelligence gathering.

---

324    Ibid.

## Conclusion

The contribution of 9/11 events to the international intelligence sharing is undoubtedly immense. Because intelligence work is a product of team effort, there are certain peculiarities common to the bureaucratic environment, such as in the case of intelligence sharing on the LRA, which help explain either success or failure. The contribution of 9/11 is seen in the US intervention in the LRA war in Uganda. The significance of the US in the trilateral intelligence sharing agreement between Uganda, South Sudan and Sudan is seen in its ability to force Sudan to join the LRA war. Through US engagement, Sudan was able to accept sharing intelligence with Uganda on the LRA, something that has been difficult before the US involvement. The US involvement led to intelligence sharing. Intelligence sharing was mainly to find the location of Joseph Kony in particular and stopping Sudan from providing logistical support to the LRA

# Chapter 10

~~~~~~~~~~~~~~~~~~~~~~~~~~~~

## The Significance of
## Intelligence Sharing Against the LRA

**Introduction**

This section discusses and examines the importance of intelligence sharing against the LRA. The intelligence sharing was taken as the last resort to defeat the LRA after the Uganda Defence Forces faced some challenges to completely put to an end the rebellion of the LRA. This brought Sudan-the main logistics supporter of the LRA, and South Sudan, that has been Uganda's strong ally for years. African Union and the US forces which were behind intelligence sharing agreement also joined in to booze the move the Uganda, Sudan and South Sudan.

## The Bangui Sub-Regional Move on LRA

Early in 2009, the foreign Ministers of the four affected countries met in Bangui and sought to consult with the AU and highlight how the LRA problem had become a regional threat[325]. When, in August 2009 and July 2010, the AU held special sessions of its Assembly in Tripoli and Kampala, respectively, the meetings, among other issues, urged the four affected countries to renew their efforts, including through military action, to neutralize the LRA and its destabilizing activities[326]. With support from the international community and friendly governments, the four set out the modalities and established the Regional Cooperation Initiative (RIC)[327]. In June 2011 at a Second Regional Ministerial meeting of the affected countries in Addis Ababa, the AU defined the strategic objective of the proposed regional initiative as "the elimination of LRA, leading to the creation of a secure and stable environment in the affected countries"[328]. Five months later in November 2011, the AU PSC declared the LRA a terrorist organization. Thus, giving more confidence to the principal participants in the sharing of intelligence against LRA namely the US, Uganda, South Sudan and Sudan.

Developing the intelligence for such purpose of defeating the LRA

---

325     Sylvester B. Maphosa. The Lord's Resistance Army: A Review of African Union Regional Efforts to Eliminate the Resistance in Central Africa, in A Comprehensive Review of African Conflicts and Regional Interventions 2016. P. 212-264

326     Letter dated 25th June 2012 from the Secretary-General Ban Ki-Moon addressed to the President of the Security Council S/2012/481 (12-38842), 6.

327     This was a shared strategy was between the AU with the UN through the US government. The US move included (a) placing the LRA as a terrorist rebellion, and (b) enacting the LRA Disarmament and Northern Uganda Recovery Act of 2009.

328     See the AU Regional Cooperation Initiative against the LRA at www.au.org. And also, letter dated 25th June 2012 from the Secretary-General Ban Ki-Moon addressed to the President of the Security Council S/2012/481 (12-38842), p. 6.

requires the need to anticipate, prevent, disrupt, or mitigate the effects of LRA attack, and also it requires the production of intelligence in a collaborative and integrated endeavour by the US, Uganda, South Sudan and a number of other agencies operating in the area. Two issues are of particular importance in understanding how US, Uganda, Sudan and South Sudan and to the large extent DRC and CAR, frame the LRA. Firstly, the political importance of domestic constituencies advocating for action against the LRA plays a significant role in determining the degree to which particular governments are able to frame the LRA in directions that are related to the security situation. In LRA affected countries, the affected populations happen to be politically marginalized due to their ethnic sympathies towards the rebels and thus in turn to not have any impact on their government's framing of the issue.

This was certainly the case in the Democratic Republic of Congo, a situation compounded by the fact that the government had much more significant threats to its territory, such as permanent instability in the east of the country more generally, which constituted a much greater menace to state authority than LRA activity in the north-east. This was also the case for other regional governments, once the scale of the LRA's operations declined. Contrary to the directly affected governments, the US government had a powerful constituency demanding action with regards to the LRA conflict. Organizations such as Invisible Children, Resolve, and Enough played a crucial role in bringing the LRA issue to the Centre of US attention. This led ultimately to a situation where the imperative of dealing with the LRA was imposed on the region by external actors, in this case the USA.

However, it had been difficult to know how genuine Sudan under al Bashir was ready to share intelligence with an open heart, since Sudan was viewed an architect of the LRA. It has been argued that the Sudanese government's constant replenishing of the LRA's supplies contributed to strengthening the movement's resilience. There have been allegations put forward that if it were not for Sudan's involvement and its support for LRA activities, the UPDF would have defeated the LRA[329]. However, it is doubtful that it was only Sudan's support that strengthened the LRA. Years after Sudan withdrew its support and allowed Uganda's forces to destroy the LRA's operational bases in South Sudan, the UPDF, the SPLA, and the US have still not been able to capture Kony, which was the main objective of intelligence sharing.

Since 2008, the United States has provided support to Ugandan led military operations to capture or kill LRA commanders, which since 2012 have been integrated into an African Union (AU) Regional Task Force against the LRA. The Obama Administration expanded US support for these operations in 2011, by deploying US military advisors to the field. In 2014, the Administration notified Congress of the deployment of US military aircraft and more personnel to provide episodic "enhanced air mobility support" to African forces[330]. The United States has also provided humanitarian aid, pursued regional diplomacy, helped to fund early-warning systems, and supported multilateral programs to demobilize and reintegrate ex-LRA combatants. The Administration has referred to these efforts as part of its broader commitment to preventing and mitigating mass atrocities. Growing

329    Ahere, J and Maina Grace. The never-ending pursuit of the Lord's Resistance Army: An analysis of the Regional Cooperative Initiative for the Elimination of the LRA. ACCORD, Issue 024, March 2013

330    Arieff, A, Blanchard, L P, and Husted, T F. The Lord's Resistance Army: The U.S. Response, September 28, 2015. www.crs.gov

US involvement may also be viewed in the context of Uganda's role as a key US security partner in East and Central Africa[331]. Thus, the fight against the LRA brought in the US, with its huge resources including advanced technology not comparable to any of the of countries collaborating in the LRA fight. The US participation was viewed to be a positive international move and blessing in the war against the LRA. Unfortunately, disparities among the US and its partners' in the fight against the LRA proved to be a serious challenge as the cooperation proceeded.

## Intelligence Sharing and National Security

The significance of the intelligence sharing is explained within the context of its contribution now and in the past. During and after the end of the Cold War, world powers namely the US, former United Soviet Socialist Republic (USSR), the UK, France and China concentrated their efforts on intelligence sharing to enhance effectiveness of international politics. This concentration was meant to make sure that their citizens and governments were safe, and that their daily activities were not interrupted. Thus, the main benchmark of intelligence sharing has been tied to the preservation of national security.

The US in particular, views the intelligence partnership as a mechanism to minimize deadly wars such as the World Wars repeating themselves between nations of different ideologies and interests. In this way, the US Departments of Defence and State in realisation of this vision, seeks to develop strong alliances across the world through military and diplomatic relations. These alliances have been proven to be useful and important to the US over the past several years. Other nations too

---

331     ibid.

have adopted the similar strategy of military and diplomatic relations to have them access to intelligence that helps them protect themselves.

For this reason, the US national security interest has long been rested upon international cooperation between intelligence services. In order to understand this, one must consider its history. As early 1940s, the posts WWII era ushered in a series of both formal and informal measures that have linked the nation's various intelligence agencies. The Central Intelligence Agency (CIA) and the National Security Agency (NSA), in the case of the US, were central in coordination of intelligence activities with their foreign counterparts.

The significance of the US intelligence sharing and cooperation featured in many situations. For example, close sharing of information between the US and the United Kingdom in 2006 thwarted a plot to destroy civilian aircraft over the Atlantic Ocean. In another example, in 2003 US and Pakistan collaborated to capture alleged 9/11 mastermind Khalid Shaykh Muhammad. Furthermore, the US has Latin America countries by providing satellite imagery in order to combat narcotics production. Still yet, another advantage has been seen in the area of US special intelligence relationships with the United Kingdom, Canada, Australia and New Zealand, which became useful for increased information sharing among these countries. This multilateral relationship, which was developed in the post years of the World War II, culminated into a high-level annual meeting that served as a platform for discussion of the various problems facing these nations.

In the case of the intelligence sharing between US, Uganda, South Sudan and Sudan on the LRA, the immediate significance was the clearance of Uganda borders free from the LRA activities and the further significance was the forcing of LRA from South Sudan territory deep into DRC and CAR. Also, this collaboration was able to achieve President Obama's policy towards the LRA. Obama's policy defined the LRA as a terror organization that must be dealt with in any form. Thus, the US has indirectly benefited from intelligence sharing relationships with Uganda, South Sudan and Sudan because it widened its effort to combat terrorism worldwide.

Therefore, the US by sharing intelligence with Uganda, and South Sudan has fulfilled its main reason behind the establishment of the CIA which was to ensure that there would be no future catastrophic resulting from security failure. It was created to avoid repetition of prior incidents such as the December 1941 Japanese surprise attack on Pearl Harbour[332]. Fortunately, George Bush's post 9/11 declaration of a war against terrorism left an opportunity for every country to define terrorism in its own terms. For example, the Uganda government define LRA a terrorist organization because of its application of terror tactics on the people of northern Uganda. Another result of Bush's strict policy on terror was that it resorted to the formation of partnerships with groups and political parties that were in opposition to their own government policies, especially those that were linked to terror organizations like Sudan. Thus, the SPLA became a partner in this arrangement.

This has not been surprising because the US has always considered its own national security paramount. This explains why the US has historically worked very closely with other countries' security organizations. even though those cooperation's are often viewed complex and require special oversight scrutiny from the oversight institutions such as the US Congress.

---

332    Gill, p. 103.

Some examples of this kind of cooperation deserve mention here. During the Afghanistan war with Soviet the US and Pakistan, along with Saudi Arabia, worked together to fund, train and equip Afghans fighting the Soviet Union in Afghanistan during the 1980s. This was facilitated by the US Inter-Services Intelligence (ISI). This nature of facilitation contributed to the development of ISI and today, the ISI remains an important and invaluable resource for the Intelligence Community helping to locate Islamic militants, Taliban operatives and top members of al-Qaeda in their diverse locations. Without assistance from the ISI, US efforts to apprehend or eliminate major terrorist threats around the world would be impossible.

The US, Uganda, South Sudan and Sudan intelligence partnership also gave kick backs. Obviously, it was observed that both US and Sudan have benefited from intelligence partnerships, especially in the US fight for counterterrorism. Available information suggest that the US and Sudanese intelligence officials worked together to track Osama bin Ladin when he resided in Khartoum during the 1990s. further and according to the press, Sudan has also occasionally assisted the US in tracking al-Qaeda operatives. Nevertheless, the US Sudan relationship has been strained because of serious and legitimate concerns about the Sudanese government's involvement in the genocide in Darfur.

**Intelligence Failure in the LRA War**

Intelligence failure has been highlighted in the past and in the present. The worst kind of intelligence failure registered in world history is a surprise attack. A surprise attack is defined as a bureaucratic neglect of responsibility, or responsibility so poorly defined or delegated that action gets lost[333]. A surprise attack is a major intelligence failure. There are notable examples of the worst surprise attacks that the world has seen. The 9/11 attack on US remains the best fresh examples of intelligence failure. In his article titled *Terrorism Early Warning and Co-Production of Counterterrorism Intelligence,* John P. Sullivan argued that contemporary terrorism is a complex phenomenon involving a range of non-state actors linked in networked organizations[334]. He further argues, "these organizations, exemplified by the global jihadi movement known as al-Qaeda, are complex non-state actors operating as transnational networks within a galaxy of like-minded networks and causing havoc among the innocent people"[335].

Correctly these groups pose security threats to countries worldwide and the collective global security is therefore urgently and without delay needed. Thus, traditional security and intelligence approaches which are focused on self must be avoided in order to achieve exclusive human security through intelligence sharing because terror groups exploit these loopholes. This holistic approach was realized in Los Angeles when it established in 1996 the Terrorism Early Warning Group (TEW) concept as a way to bridge the gaps in traditional intelligence and security structures. The TEW embraces a networked approach to intelligence fusion and directs its efforts toward intelligence support to regional law enforcement, fire and health agencies involved

---

333     Schelling, T. (1962). "Forward" in B. Wohlstetter, Pearl Harbor: Warning & Decision. Palo Alto, CA: Stanford Univ. Press.

334     John P. Sullivan. Terrorism Early Warning and Co-Production of Counterterrorism Intelligence. Canadian Association for Security and Intelligence Studies CASIS 20th Anniversary International Conference Montreal, Quebec, Canada, 21 October 2005. p.1.

335     Ibid.

in the prevention and response to terrorist acts. This was not the first time US opted to use intelligence. Effective intelligence sharing was a crucial element of pacification strategy that the US adopted in 1967 with both South Vietnam and the US contributing valuable intelligence to this effort that the other found to be too costly and difficult to collect[336].

Causes of intelligence failure are multifaceted and varied. The joint congressional inquiry into the causes of the Pearl Harbor failure concluded that lack of coordination among the different government departments producing intelligence and the lack of integration of their product, were the major causes of intelligence failure recognized at that time[337]. Also misunderstanding of the situation that leads a government and military forces to take actions which are inappropriate and counterproductive to its own interests, has been identified as another cause of intelligence failure[338]. However, the causes of intelligence failure in the US, Uganda, South Sudan and Sudan declared war on the LRA were basically about the lack and openness of Sudan to provide accurate information about the LRA. In fact, Sudan was accused of leaking information on the movement of forces to LRA leaders. The other cause of intelligence failure was the subordination of SPLA throughout the operations. This subordination was noted in the US military and financial support. US concentrated its support to

---

336    James Igoe Walsh. Intelligence Sharing for Counter-Insurgency. Defense and Security Analysis Vol. 24, No. 3, pp281-301, September 2008.

337    Davies 1, P. H. (2004). Intelligence culture and intelligence failure in Britain and the United States. Cambridge Review of International Affairs, 17(3), 495-520.

338    Hatlebrekke, K. A., & Smith, M. L. (2010). Towards a new theory of intelligence failure? The impact of cognitive closure and discourse failure. Intelligence and national security, 25(2), 147-182.

UPDF, who were not familiar with South Sudan terrain and were seen as alien forces, thus bringing into question the cooperation with civil population.

## Intelligence Sharing Failure at International Level

The beneficial nature of intelligence sharing has in the last couple of years compelled states as well organizations to increase their willingness to collectively share intelligence. The desire to open up exchange of information and share their knowledge to address threats, vulnerabilities, and incidents. This intelligence exchange is part of states mitigation strategies that result from states' growing need to protect themselves against today's sophisticated terrorists' attacks. This strategy can be interstate, inter regional or intercontinental. In 2013, Whitelock reported that the US military is expanding its secret intelligence operations across Africa, establishing a network of small air bases to spy on terrorist hideouts, from the fringes of the Sahara to jungle terrain along the equator, according to documents and people involved in the project[339].

In Africa, reportedly, formal talks among national intelligence services began as early as 1992, "when African leaders meeting in Dakar, Senegal, first raised concerns about growing radicalization and extremism on the continent"[340]. The rationale behind intelligence, as already alluded elsewhere in this work, emanates from increasing worries about the spread of terror groups such as Boko Haram, an Islamist

---

339    Whitlock, C., 2012. US expands secret intelligence operations in Africa. The Washington Post, 13.

340    Cline, L.E., 2016. African Regional Intelligence Cooperation: Problems and Prospects. International Journal of Intelligence and Counterintelligence, 29(3), pp.447-469.

group in Nigeria blamed for a rash of bombings there. In Somalia, US forces were orchestrating a regional intervention to target al-Shabab. In Central Africa, about 100 American Special Operations troops were deployed to help coordinate the hunt for Joseph Kony. However, the results of the American surveillance missions are shrouded in secrecy.[341] Although the US military has launched airstrikes and raids in Somalia, they generally limit their involvement to sharing intelligence with allied African forces so they can attack terrorist camps on their own territory.

In evaluating intelligence sharing between the United States and Britain and West Germany during the early Cold War, Walsh has argued that intelligence sharing among countries with different technical capabilities or asymmetries and local knowledge is particularly valuable for countering terrorism, transnational organized crime, and the proliferation of weapons of mass destruction[342]. States such as US, Uganda and South Sudan which participate in intelligence sharing arrangements, must balance the benefits of more and better intelligence against the possibility that their partners will withhold or distort the information they share, or will pass along to others the information they receive[343]. Participants can balance these benefits and risks by introducing elements of hierarchy into their sharing agreements. Hierarchical arrangements allow a state to monitor more effectively for defection and to reassure others of its own commitments to the terms of their sharing arrangement. In the case of the US and Pakistan,

---

341      opcit

342      Walsh, J.I., 2007. Defection and hierarchy in international intelligence sharing. Journal of Public Policy, pp.151-181.

343      Ibid.

intelligence failure was attributed to leakage of information by elements of the Pakistan's military who were found to be sympathetic to the Taliban and other militants. These suspicions have at times strained the US Pakistan security relationship.

Many theorists try to understand why intelligence fails. According to James Walsh, one of the obstacles in terms of sharing international intelligence, is gathering intelligence in an aggressive way[344]. Therefore, to ensure international cooperation and intelligence sharing, there should not be an aggressive attitude in intelligence activities; there should be specific ethics, norms and a value system. In addition to this, according to Walsh, it is useful to use a hierarchy mechanism to establish international intelligence sharing system. On his part Otwin Marenin pointed out that international intelligence collaboration should involve information sharing directed to specific operations[345], that is to say, information sharing can be enabled within a limited area.

After the 9/11 attack, intelligence failure was explained using three main conspiracy theories. First it was suggested that the CIA did it; secondly, Israeli intelligence did it; and thirdly, the Arabs did it, and the CIA let it happen[346]. However, what is evident and clear later through investigation was the catastrophic weaknesses relating to human error exposed with passport & visa system, which indicated that background checks were not done. It was suggested that the FBI operated on "dumb" pencil-and-paper mode, while the CIA didn't have

---

344    Beren, F., Can International Intelligence Sharing System Be Established For Global Security.

345    Marenin, O., 2006. Democratic oversight and border management: Principles, complexity and agency interests. Borders and Security Governance. Vienna and Geneva: LIT Verlag/DCAF, pp.17-40.

346    See Hart 2003.

enough linguists and translators. Furthermore, it was suggested that airport security didn't know how to perform technicalities such as screening, and that there was too much distrust of sharing secrets, and civilian agencies didn't know how to handle military threats, while military agencies didn't know how to handle law enforcement threats.

In 2004, a Commission to look into the issues surrounding 11 September terrorists attack on the Twin Towers was constituted. This Commission in its final executive report **specified numerous** intelligence failures. These failures included neglect to follow up closely those potential terrorists who were on watch list. The list included future hijackers such as Hazmi and Mihdhar. As a result, these terrorists were not trailed after they traveled to Bangkok. Furthermore, it was found out there were not sharing of information that linked individuals in the *Cole* attack which was linked to Mihdhar. Additionally, it was highlighted that intelligence personnel were not taking adequate steps on time to find those suspects, namely Mihdhar and Hazmi, when they were actually inside the United States.

The report suggested that intelligence officers were not linking the arrest of Zacarias Moussaoui, described as interested in flight training for the purpose of using an airplane in a terrorist act, to heightened indications of attack. Again, it was underlined that intelligence officers were not able to discover false statements on visa applications by the Hamburg cell; and were an unable to recognize passports manipulated in a fraudulent manner by the Hamburg cell; hence, they failed to expand no-fly lists to include names from terrorist watchlists. More importantly, the intelligence officers were so ignorant by not searching airline passengers identified by the computer-based CAPPS screening system and finally, to make it worse, they failed to harden aircraft cockpit doors or taking other measures as preparation for the possibility of suicide hijackings.

However, former CIA expert, Michael Scheuer, who authored two books on intelligence: *Imperial Hubris and* **Through Our Enemies› Eyes**, challenged the 9/11 Commission Report, arguing that the failure was erroneously based on a portrayal of the problem resulting from budgetary, structural, and organizational issues[347]. In a letter to the House and Senate Intelligence Committees, Scheuer laid out the "Ten Steps How *Not* **to Catch a Terrorist,**[348]" Despite ample evidence concerning the ability and the intention of the terrorists to launch an attack on the US, the intelligence agencies involved failed to provide a timely and accurate warning. However, the 9/11 surprise attack were not the only one of its kind. Along with the German attack against the Soviet Union in June 1941 and the Japanese attack on Pearl Harbor in December 1941, the coordinated Egyptian-Syrian attack against Israel on Yom Kippur, 6 October 1973, are all considered a classic examples of a successful surprise attack and costly intelligence failure.[349]

Although past and historical security events around the globe have been identified to have connections internally and internationally with potential security issues (for they become factors of causation and effect)[350], existing numerous sources have also analyzed general-

---

347    Scheuer, M. (2002). Through Our enemies' eyes: Osama bin Laden, radical Islam, and the future of America. Potomac. Also, see Scheuer, M. (2004). Imperial hubris: why the West is losing the war on terror. Potomac Books.

348    For details see Scheuer, M. How Not to Catch a Terrorist? The Atlantic. December 2004 Issue www.theatlantic.com

349    Bar-Joseph, U. Arie W. Kruglanski (1995) Intelligence Failure and Need for Cognitive Closure: On the Psychology of the Yom Kippur Surprise.Political Psychology, Vol. 24, No. 1, 2003.

350    Carr, E. (1961). What is History? NY: Vintage Books.

ly reasons attributing to that, and have attributed the main cause as certain tendencies which are inherent in most bureaucratic security institutions[351]. As Bar-Joseph and U. Arie W. Kruglanski point out:

> *Newly available evidence to shed light on the circumstances and causes of the 6 October 1973 Yom Kippur surprise attack of Egyptian and Syrian forces on Israeli positions at the Suez Canal and the Golan Heights suggests that an important circumstance that accounts for the surprise effect these actions managed to produce, despite ample warning signs, is traceable to a high need for cognitive closure among major figures in the Israeli intelligence establishment. Such a need may have prompted leading intelligence analysts to "freeze" on the conventional wisdom that an attack was unlikely and to become impervious to information suggesting that it was imminent. The discussion considers the psychological forces affecting intelligence operations in predicting the initiation of hostile enemy activities, and it describes possible avenues of dealing with the psychological impediments to open-mindedness that may pervasively characterize such circumstances[352]*

However, in a real sense, intelligence failure has been linked at some point to overestimation. This was probably the cause of intelligence failure during the pursuit of LRA leader Joseph Kony. The overestimation contributed to the inability to capture or kill Joseph Kony. US, Uganda and South Sudan forces overestimated their capacities and underestimated the LRA. This is perhaps the most common reason for

---

351    See Laqueur, Walter. (1985). A World of Secrets: Uses & Limits of Intelligence. NY: Basic, Lowenthal, M. (2003). Intelligence: From Secrets to Policy, 2e. Washington D.C.: CQ Press.

352    Bar-Joseph, U. Arie W. Kruglanski.

failure. Leaving it uncorrected can lead to the continuation of error for a long time. Importance lessons can be learned from Cold War experiences. During the Cold War period, the US consistently overestimated the missile gap between the US and Soviet Union. Critics of the Iraq War believe that overestimation was the main error that happened in estimating Saddam Hussein's capabilities. Over confidence occurs when one side is so confident of its ability that it projects its reasoning onto the other side and believes that since it would not do something itself, neither will the other side. The classic case is the Yom Kippur war of October 1973, although the whole Cold War was characterized by this.

Also, it has been noted that, underestimation **occurs when intelligence or political leadership seems unwilling to be receptive to warnings, or completely misread**s the enemy's intentions. A classic example is drawn from Stalin in 1941, who turned a blind eye to the possibility of Hitler invading Russia, despite the fact that the British and Americans tried to tip him off with that information. This clearly indicates how lack of trust in foreign intelligence may also be a reason explaining intelligence failure. Another account for intelligence failure has been subordination of intelligence to policy. Intelligence subordination happens when judgments are made to produce results that superiors would want to hear instead of what the evidence would suggest. Because of commonality of this, intelligence subordination has become the most widely discussed and analyzed type of intelligence failure because of the error and bias associated with it. In relation to this, it is argued that with 9/11, there was possibility that a "hands-off" policy towards Saudi Arabia which interfered with intelligence over the hijackers, who were mostly from Saudi Arabia.

Another cause of intelligence failure has been associated with lack of communication. The lack of a centralized communication office often creates the problem of lack of communication; however, it more typically results from amalgamation of different officials from different agencies with different rules, different security clearances, and different procedures on who and how they communicate. It also occurs when there are too few analysts who only work on-the-fly for different agencies and don't have full-time intelligence responsibilities. There is also unavailability of information. It is recognised that regulations and bureaucratic jealousies are sometimes the cause of this, but the most common problem involves restrictions on the circulation of sensitive information. When there is virtually no intelligence at all, this is called something else: ignorance.

Before the terrorists of 9/11 could bring down the Twin Towers there was a 'received opinion' **sometimes called «conventional wisdom»** which consists of assertions and opinions that are generally regarded in a favourable light, but have never been sufficiently investigated. Sometimes the people in a bureaucracy are forced to make "best guesses" on the basis of limited information. 'Mirror imaging', is technically defined as "the judging of unfamiliar situations on the basis of familiar ones," but most often involves assessing a threat by analogy to what the government would do in a similar position. There is also the problem of having too many areas of specialist intelligence.

During the intelligence sharing exercise towards LRA, an element of complacency **was observed. Complacency happens when the enemy is known to do something, though there is not surety in information or when,** so therefore nothing is done. The classic example is the British who did nothing in the weeks leading up to the

Falkland War of 1982. A modern example is the way the international community sat on the side-lines during the Rwanda massacre. There's a tendency in some circles to just let things run their course. And finally, there was intelligence failure to connect the dots. The connections between bits of intelligence were not put together to make a coherent whole. It is most easily observed in hindsight and is perhaps the main cause behind why the 9/11 attacks caught American officials by surprise.

As much as the agreement on intelligence sharing against LRA succeeded to bring LRA atrocities theoretically to an end, the fact that Joseph Kony, the LRA leader remained at large, signifies the failure of intelligence. Originally, the plan was that Joseph Kony must be found alive or dead. According to Richard Betts, the most crucial mistakes leading to intelligence failure have seldom been made by the collectors of raw information, occasionally by professionals who produce finished analysis, but most often by the decision-makers who consume the products of intelligence services.[353]

It was impossible for the international community to turn away while watching the merciless LRA killing of innocent armed and non-combatants. International norms require that war is fought within the limit of Laws of Armed Conflict (LOAC). Throughout its nearly four decades of existence, the LRA has violently waged war with no distinction between military and civilian targets. Based on this, it is undeniable that the LRA conduct of war fell below the international required standards of fighting a just war, resulting in the LRA being perceived as evil. John Prendergast wrote that "the fate of a war

---

353     Betts, R.K. 1978. 'Analysis, War, and Decision: Why intelligence failures are inevitable'. World Politics, 31,1, pp. 61-89.

that has crossed three international borders, displaced nearly two million people, and created the highest child abduction rate in the world hinges on the fate of one man: Joseph Kony, the notorious leader of the rebel Lord's Resistance Army (LRA)"[354]. This, in essence, meant that the LRA's evil acts should not be left unpunished, therefore, the need for the international community to urgently address the LRA rebellion, whose deeds surpassed toleration - especially when those acts were transnational and cross cutting - it was urgent to help the Uganda government in many ways, including by sharing intelligence against the LRA. This was in response of the call by the Uganda government, who on its part, reported LRA atrocities and called for international help.

Reacting to these calls by the Uganda Government, the US and African Union (AU) both concluded the need to collectively fight the LRA so that its barbaric atrocities were brought to a stop. Therefore, the process to eliminate the LRA was launched following a Special Session of the Assembly of the AU on the Consideration and Resolution of Conflicts in Africa, held in Tripoli, Libya on 31 August 2009. That session adopted the Tripoli Declaration on Elimination of Conflict and the Promotion of Sustainable Peace in Africa, with a plan of Action towards that end[355]. The Ordinary Session of the AU Assembly, held in Kampala, Uganda, from 25-27 July 2010, requested the AU commission, within, the framework of the Tripoli Plan of Action, to organize action-oriented consultations between the LRA affected countries, and all other interested parties, with a view to facilitating coordinated

---

354    John Prendergast, WHAT TO DO ABOUT JOSEPH KONY. ENOUGH Strategy Paper #8 October 2007. Available at https://www.americanprogress.org/wp-content/uploads/issues/2007/10/pdf/kony_report.pdf

355    Report of the Deputy Commander AU RTF to A/CDF for Ops, Trg and Int.

regional action against the LRA and its threat. As a follow-up, the AU undertook extensive consultations with the countries affected by the activities of the LRA namely Uganda, South Sudan, DRC, and CAR, in addition to other countries with great interest such as the US. Two regional ministerial consultative meetings were also held on the LRA problem; the first in Bangui, in October 2010 and the second in Addis Ababa, in June 2011. Throughout the consultations, the affected countries expressed commitment and determination to coordinate efforts to eliminate the LRA and its terrorist activities from the region[356]. Another AU Summit held in Malabo, Equatorial Guinea, from 30 June -1 July 2011, endorsed the conclusions of the 2nd regional ministerial meeting of June 2011, and requested the AU Peace and Security Council (PSC) to authorize the proposed operation in all its components, including the Regional Task Force (RTF), the Joint Operations Centre (JOC) and the Joint Coordination Mechanism (JCM). It also requested the UN and other AU partners, to support this initiative aimed at protecting affected civilian populations; consequently, on 22 November 2011, the PSC authorized the Regional Cooperation Initiative for elimination of the Lord's Resistance Army, as an AU initiative, supported by the international community.

The mandate of the Regional Cooperation Initiative for the elimination of the LRA included: firstly, strengthening the operational capabilities of the countries affected by the atrocities of the LRA; secondly, creating an environment conducive to the stabilization of the affected areas; and finally, facilitating the delivery of humanitarian aid to affected areas. In order to implement the mandates given above, the AU PSC approved the establishment of the different components of

---

356    Report of the Deputy Commander AU RTF to A/CDF for Ops, Trg and Int.

the Regional Cooperation Initiative for the elimination of LRA. The first component is the Joint Coordinating Mechanism (RCM), chaired by the AU Commissioner for Peace and Security, and comprising the ministers of Defence of the affected countries at the strategic level to coordinate the efforts of the AU and the affected countries with the support of international partners. And the second component is RTF composed of units, provided by the affected countries with strength of 5000 soldiers, to be based in DRC, CAR and South Sudan.

The elimination of the LRA included the elimination of its top command, especially Joseph Kony and his deputies. This has been clear in the conception of the operations' mission to kill, capture, or force defections of LRA members, with spectrum combat operations across South Sudan, CAR and DRC. This resulted into the killing of three LRA commanders, and one on trial at the Hague, with Joseph Kony remaining at large. More importantly, as additional measures Joseph Kony, its top command and its organization, were targeted by sanctions. The use of targeted sanctions as a central instrument to address challenges to international peace and security has been a defining feature of UN Security Council practice since the end of the Cold War[357].

In this regards, international and regional action against the LRA is viewed as an international strategy championed by the US and its allies AU meant to bring an end to LRA human rights abuses in northern Uganda and South Sudan. The result was the marching in of the Ugandan army and the SPLA, backed by US and AU to search

357    Thomas J. Biersteker, Sue E. Eckert and Marcos Tourinho (2016). Targeted Sanctions: The Impacts and Effectiveness of United Nations Action. Cambridge University Press. P. 1.

for Joseph Kony and his colleagues in the equatorial bushes of South Sudan, DRC and CAR. Encouraging these moves were reports displaying LRA horrific abuses, no longer secret in the eyes of international community. On March 28, 2002, Human Rights Watch (HRW) report released a horrific account of the massacre of over 321 civilians by the LRA, alongside the abduction of over 250 individuals. The report revealed barbaric actions in which innocent civilians were hacked to death by machetes or clubbed to death by axes. The HRW called for an international, comprehensive strategy to apprehend Joseph Kony and the LRA. It also called for the disarmament of the LRA. Aaron Edwards notes that: "Regardless of whether they are part of a state's armed forces or a guerrilla groups, lawful combatants must all adhere to the law of armed conflict, which consists of four basic principles namely distinction, humanity, military necessity, and proportionality for application of armed forces."[358]

The US responded by adopting the Northern Uganda Recovery Act (2002). This Congressional Act allowed the United States to provide political, economic, military, and intelligence support for multilateral efforts that seek to protect civilians from the LRA and his leader Joseph Kony. Also, the Act called on the United States to respond to the humanitarian needs of those affected by LRA activities, especially the needs of people in northern Uganda. On its part, the Senate and the house of Representatives unanimously passed the bill. In relation to this, Brownback, a promoter of anti-malaria programs in Africa, who visited Uganda in 2004, urged the United States and the international community to work closely to establish lasting peace in

---

358    For details of these principles see Aaron Edwards (2017). Strategy in War and Peace: A Critical Introduction. Edinburgh University Press. P. 105.

northern Uganda and to bring Joseph Kony and the Lord's Resistance Army to justice.

In December 2003, President Museveni requested the ICC to investigate the LRA for committing massive killings, abductions, kidnapping of children, and other inhumane treatments against the civilian population. This resulted in the indictment of 4 top LRA leaders, namely, Joseph Kony, Vincent Otti Odhiambo and Dominic Ongwen. It is worth noting that the creation of the ICC in July 2002 was to help stop the worst crimes against humanity inflicted by governments or rebel forces against their own citizens, with the intention that criminal justice would contribute to peace-making. In this sense, the ICC became relevant to the Uganda case.

At the East African Community (EAC) level, counter measures were put in place and included: the Anti-terrorism Act (2002); the Amnesty Law (2000) to encourage those who wanted to renounce membership with the LRA; the Anti-money laundering bill; and the EAC Protocol on Peace and Security to jointly address the issue of security in the region including terrorism. The United Nations Mission in the Sudan (UNMIS) identified the LRA was making it difficult for the returning population of South Sudan to settle in their new areas."[359] On 17 August 2005, the returnees in some areas of Eastern Equatoria wrote a letter to the President of the GoSS, on the threats posed by the LRA to their lives and the success of the CPA. On 24 August, Salva Kiir - GoSS President advised the LRA to seek political settlement rather than continue fighting. He indicated that the GoSS would otherwise be compelled to forcefully remove them from South

---

359     UNMIS Briefing note on the status of implementation of the comprehensive peace agreement (CPA) in the Sudan, 29 September 2005.

Sudan. Kiir also offered to meet with the head of the LRA personally, at a time and place convenient to the latter. The result was the birth of the Juba Peace Talks.

All nations of the globe have engaged in intelligence collection throughout their time of existence. However, it has been argued that contemporary intelligence networks date back to the Cold War.[360] The intention behind sharing information is to coordinate operations in order to address mutual problems such as terrorism, human trafficking, pandemic diseases, environmental concerns, and so forth. After the fall of the Soviet Union and the world became unipolar under the leadership of the United States, agencies found that, in a globalized world, they were increasingly called upon to combat problems spanning borders, such as drug and human trafficking.[361] Thus, the history of intelligence sharing or information exchange is long, rich and rewarding. For instance, coalition forces in Afghanistan formed networks to exchange information about military operations; financial regulators established formal links to enforce sanctions; and law enforcement agents institutionalized more frequent contact with other services.[362]

The impact of intelligence sharing on LRA has been productive in the sense that the AU RTF managed to clear the LRA out of Uganda and South Sudan borders pushing it to the DRC, CAR and Sudan borders. It also reduced the original LRA leadership by killing three commanders and forced one of them to surrender where he was handed

---

360     see Richard Aldrich, (2001), the hidden hand: Britain, America and Cold War Secret Intelligence 8–9.

361     Elizabeth Sepper, "Democracy, Human Rights, and Intelligence Sharing" Texas International Law Journal Vol. 46:151.

362     Ibid.

over to the ICC for trial at the Hague, Netherlands. Of the original brutal commanders, only Joseph Kony is still alive and is in hiding in unestablished location sometimes said to be in Sudan's Darfur region or in the Central Africa Republic.

## The Departure of the African Union Force

The reduction of the LRA capabilities was marked by the closure of the African Union Rapid Response Task Force (AU RTF) facilities and its redeployment out of South Sudan. This came as a result of the 6th Ministerial Joint Coordination Mechanism meeting in Addis Ababa, Ethiopia during the JCM of the Regional Cooperation Initiative for the elimination of LRA that took place in Addis Ababa on 30 March 2017. In this meeting, the US announced that its special forces would cease operations on 26 April 2017. Similarly, the Uganda government announced that its UPDF will withdraw from RTF immediately after the withdrawal of the US special forces. The withdraw of the US from pursuit of the LRA was a major setback in the regional fight against the LRA. As the US announced its pullback, the Uganda government also announced its troops' withdrawal and the SPLA in South Sudan eventually had to abandon the fight. In fact, South Sudan pointed out clearly that it would not be able to host the Headquarters of the RTF without the support of US special forces and UPDF[363]. This clearly underlines how the intelligence sharing among the unequal's can be constrained when the strongest party leaves. The departure of the US meant many things to the collaboration. It meant withdrawal of financial support and withdrawal of technology of intelligence gathering. US went, taking with it its money and its technology, therefore

---

363     Report of the AU RTF Deputy Commander to A/CDF Ops, Trg & Intel.

leaving the AU, Uganda, South Sudan, DRC and CAR to hope for the best. Without US presence in the fight against LRA, all achievements recorded would not have been possible. What has been achieved has been achieved but the mission to eliminate the LRA is unaccomplished. War and the fear of war have always necessitated the need for intelligence sharing among the nations. The Secret Service Bureau was established at a time of heightened and intensifying international rivalries when British strategic policymakers were becoming especially concerned about the challenge of an aggressive, ambitious, imperial Germany.[364]

Intelligence sharing involves collection, analysis and management. Collection involves intelligence sharing among countries with different technical capabilities and therefore local knowledge is particularly valuable for countering terrorism, transnational organized crime, and the proliferation of weapons of mass destruction. States participating in intelligence sharing arrangements must balance the benefits of more and better intelligence against the possibility that their partners will withhold or distort the information they share or will pass along to others the information they receive. Participants can balance these benefits and risks by introducing elements of prioritization into their sharing agreements. Prioritization arrangements allow a country to monitor more effectively for defection and to reassure others of its own commitments to the terms of their sharing arrangement.

According to the information gathered in the field, the immediate contribution of this intelligence sharing arrangement has been the elimination of the LRA activities in northern Uganda and Eastern

---

364    Keith Jeffery (2010). MI6: The History of the Secret Intelligence Service 1909-1949. Bloomsbury. P. 3.

Equatoria. This claim surfaced in the interviews carried out among the SPLA and the UPDF officers who participated in the LRA operations. The interviews indicated that as soon as intelligence sharing agreement came into effect, LRA activities in northern Uganda and the neighbouring Eastern Equatoria State of South Sudan, sharply dropped. Indeed, the period which followed the launched of joint SPLA-UPDF operations on the LRA, following the collapse of Juba Peace Talks, saw an incredible drop of LRA attacks in Uganda and Eastern Equatoria, with most of the LRA activities being concentrated on the South Sudan Western Equatoria border with DRC and CAR.

Although the LRA has indeed been forced out of Uganda and South Sudan through intelligence sharing arrangements which resulted into successful operations against it, other factors such as good logistics, political changes in Sudan which manifested itself in full independence of South Sudan, internal divisions within the LRA ranks and files, were also significant in the fight against the LRA. The total relocation of the LRA from Uganda and South Sudan is undoubtedly a major victory for Sudan and South Sudan in their fight against the LRA. However, the LRA's long dormancy at least from military perspective may be considered a tactical withdrawal. In military strategy, withdrawal is never considered a defeat. It is part and partial of the military game. Indeed, history has shown that rebel movements can disappear and reappear after being forgotten for so long. The EPDF, EPRDF, ELF to just mention few, are clear examples of this kind.

Some of the failures of intelligence observed during the LRA search operations were related to: firstly, gathering of intelligence information. The first stage of intelligence process is gathering of information. The rationale behind intelligence gathering is that it helps leaders in their decision making. Intelligence can be gathered any time, during peace or war, for purposes of decision-making. Information collection therefore forms an essential part of the whole process of intelligence sharing. Despite this however, it appears that the information gathered on the LRA was uncoordinated. The US, with its advanced technology, was confident beyond doubt that it undermined information collected by the SPLA intelligence officers, not knowing that the SPLA had privilege of being indigenous and therefore held local trust.

The second, is the methods and challenges of intelligence gathering. The gathering of intelligence as part of counter LRA operations involved a number of methods. This includes human intelligence. Human intelligence consists of professional intelligence personnel, paid informers, and volunteers. Of the intelligence officers interviewed, the prevalent view was that the SPLA majorly relied on information from the communities, paid informers and above all, from military intelligence personnel who ventured into rebel bases. Intelligence personnel from both UPDF and the SPLA located and identified the LRA positions and their activities. Also, local informants were used to alert the UPDF and the SPLA about the movement of the LRA and their commanders. Intelligence officers from both the UPDF and the SPLA carried out joint intelligence missions intended to locate the LRA hideouts. These joint intelligence-gathering missions were fruitful in the anti-LRA fight. The LRA high leadership members and forces have been considerably reduced in numbers with its forces reduced from thousands to small pockets of approximately only 200 men since the beginning of intelligence sharing operations. Also, it resulted in the LRA being pushed away from its original areas of operations in northern Uganda and southern Sudan, deep into the CAR[365]. An anonymous interviewee revealed that in March 2010, Ugandan intelligence officers followed closely Kony's movement into the southern Darfur region of Sudan, hoping to receive support from his former benefactor, the Khartoum government. On his way back into the CAR, where the bulk of his forces were, he was intercepted on the way, leaving his fighters scattered with a high death toll[366]. A content analysis of documents also revealed that the Ugandan army, through the use of human intelligence identified, killed and captured more of Kony's strong and faithful officers. The general evaluation is that the exchange of intelligence and information between the allied anti LRA forces (namely the US, Uganda, and South Sudan) is satisfactory but more effort is needed, encouraged, developed and supported so that it grows and maintains and is used for future operations. If this intelligence sharing is carefully handled, it can lead to confidence building among these countries and eventually to good neighbourly relationships.

There also exists communications and signal intelligence. Where the SPLA majorly relied on information from the communities through well-wishers, paid informers and military intelligence personnel who ventured into rebel bases, the UPDF mainly intercepted telephone and radio communications between rebel units. This was a prevalent view from the interviews carried out amongst intelligence officers of the SPLA and the UPDF[367]. But more importantly, the interviewees felt

---

365    Before intelligence sharing agreement was concluded the LRA bases were all concentrated along Uganda South Sudan borders.

366    Interview with an anonymous interviewee, Juba Southern Sudan, June 21, 2010.

367    The author is grateful the SPLA and UPDF officers interviewed.

that both human and signal intelligence complimented one another.

A content analysis of a report by Africa Crisis Group reveals that the audio intercept capability given by the US made it easy for the UPDF to intercept the LRA communications[368]. According to this document, the US military through African Command (Africom) provides communications, logistical and intelligence support for the UPDF in its pursuit of the LRA. Through technical intelligence the UPDF and the SPLA were able to follow communications between LRA commanders. These communications were further analysed and used to locate the LRA positions, followed by military strikes. As a result, the LRA planned operations were obstructed or pre-empted, leading to the killing of their top commanders. As the document revealed, technical intelligence was more reliable than human intelligence in terms of intelligence accuracy. However, two SPLA senior intelligence officers interviewed, namely Maj. Gen. Lat Zakaria and Brig. Mac Paul, perceived it differently. They felt that communication/signal and human intelligence complimented each other with different ability. As they revealed during interviews, the SPLA relied very much on UPDF as far as human intelligence was concerned, and UPDF also relied on SPLA for human intelligence.[369] Brig. Riak Jeroboam of the SPLA MI asserted that although intelligence was shared by the UPDF and the SPLA, much of the accurate information has came from the UPDF due to the fact that it had the latest interception equipment, such as satellite phones which the SPLA lacked has supported this[370].

---

368     Crisis Group Africa Report No. 156, 28 April 2010.

369     Interview with Maj. Gen. Lat Zakaria and Brig. Mac Paul, Juba 20 July 2010.

370     Interview with Brig. Riak Jeroboam, May 26, 2010.

Another important aspect of communication intelligence besides the strategy of tracking down the LRA groups and engaging them militarily with the aim of capturing or killing fighters (mostly Joseph Kony and his deputies) - and liberating abducted civilians; a communication campaign to encourage fighters to defect was also applied, a content analysis of the document revealed. Former LRA combatants made radio appeals to their former brothers-in-arms to accept amnesty and reintegration into either the army or civilian life.[371] Ground troops also left leaflets where the LRA was active, showing pictures of former commanders and written calls to disarm[372]. These two approaches were intended to be complementary, the stick and a carrot - the greater the military pressure, the more attractive the prospect of disarmament.

However, having been successful as it is, there is yet numerous criticism leveled against Sudan and Uganda and the International Sharing Agreement. Firstly, competence/skill of intelligence personnel. Based on interviews, participants' observations and documentary reviews, it is obvious that the shared intelligence was not as accurate as it was originally supposed. Intelligence information was not delivered on time or not at all. Of the intelligence officers interviewed, many felt that satisfactory exchange of intelligence on the LRA took place between South Sudan and Uganda forces, but not between SAF and the UPDF. The initial agreement of intelligence sharing was between SAF and UPDF. As a key member in the intelligence agreement Government of Sudan (GoS) has a role to play. Lack of full commitment to intelligence exchange between SAF and UPDF indicates a loophole in

371     Ibid.
372     Ibid.

the intelligence sharing agreement and confirms the allegation that Government of Sudan (GoS) still supports the LRA, despite being part of counter operations.

Lack of competence and skill of intelligence officers has contributed to the way the intelligence information itself been shared. The method employed has not actually been compatible with intelligence systems. This takes the form of joint meetings, frequent communications either by use of telephone, letters or joint reconnaissance missions. An example of this kind of joint meeting took place at Arua, Uganda on the 7th February 2007[373]. Although such meetings regularly convene and although there has been good intention in conducting them there hasbeen problem in implementation of what is actually shared. Also attributed to lack of skill is the way the intelligence missions were arranged. Face to face interviews and content analysis of personal notes documents of intelligence officers of the SPLA revealed that the SPLA and the UPDF intelligence officers moved together on many occasions to collect information and to spot the LRA movements. SPLA intelligence Capt. Abraham Chol Kuany, in one of his reports, acknowledged that they were assigned to conduct a joint intelligence reconnaissance (RECCE) and the aim of their mission "was to organize a fighting RECCE force composed of the SPLA and the UPDF"[374].' However, there is a distinction between intelligence gathering/collection and implementation of the intelligence gathered. This has been lacking with the officers interviewed. Better intelligence is based on skill intelligence personnel has. The UPDF and the SPLA did not have enough intelligence officers with enough skill of collection

---

373    From the letter of the SPLA COGS, 4th February 2007.

374    Extracted from MI report by Capt. Abraham Chol Kuany, Oct 25,2009.

and analysis. This was a prevalent view during interviews: that intelligence officers, especially members of the South Sudanese army (SPLA) urgently require training in intelligence collection and information gathering, and intelligence analysis. They also require training on how to use modern intelligence equipment such as satellite phones, GPS, etc.

The interviews conducted in Uganda with several UPDF officers, including the CDF of Uganda and his chief of military intelligence, indicated satisfaction with the contribution of intelligence sharing towards the LRA fight. Specifically, they concurred on the reduction of the LRA forces from the initial force strength of about 5000 to the current force of about only 300[375.] However, this claim cannot be confirmed since there was no access to LRA areas. It is difficult to confirm how the UPDF arrived at this number. It is likely that the numbers given as LRA strength were meant to meet media needs. The method employed by the UPDF and the SPLA, which composed of joint intelligence missions and joint military operation missions can be viewed as not following the recognised methods of intelligence collection. It is hard to differentiate them from normal combat missions. Such kinds of operations were launched in December 2008 by the UPDF with assistance from the SPLA. A content analysis of a document on intelligence sharing supports this claim. It reveals that at the signing of the protocol on the 4th March 2002, LRA strength was estimated at about 5000 fighting men, divided into five infantry brigades namely; Sima Bde (700), Stocree Bde (750), Gilva Bde (600), Trinkle Bde (850) and the Control Altar Bde (1000).[376]

---

375    Interview with Gen. Aronda and Brigadier Mugira, 19th May, 2010, Kampala.

376    Extracted from David Pulkol briefing notes, 29th March 2002.

During an interview with Gen. Aronda, the then chief of Defence Forces of Uganda, he indicated that the current LRA effective combat strength was estimated to be between 300 and below[377]. However, this claim is indisputable since numbers can be used for political reasons. An interview with UPDF MI officer, who did not want to be identified, suggested that the current LRA activities do not match with the number suggested by the CDF[378]. Uganda's participants at the border security meeting suggested the capability of the LRA to have greatly diminished during a presentation on "terrorism, and border security challenges in Uganda" held in Kampala Uganda on June 24-26, 2020, attended by East African countries and the SPLA.[379]

At several security meetings the researcher attended in South Sudan, the prevalent view was that the impact of intelligence sharing has been very great and effective specially in eliminating elements of the LRA leadership. So far 6 top commanders of the LRA, namely Vincent Otti, Bok Abudema, Ocan Bunia, Nixman Oryang (captured), and Charles Arop (captured), were either died/killed or captured since the deal was signed[380]. Arising from interviews with the same respondents it suggested that the LRA is not only reduced in manpower, but also the areas of operations. Since the deal was reached, the LRA has been pushed away from its stronghold areas in northern Uganda and along the South Sudan – Uganda Border, deep into the South Sudan and further into South Sudan, DRC and the CAR common borders. As a result of this push, a significant reduction in civilian causalities

---

377    Interview with Gen. Aronda, 19th May 2010, Kampala.

378    Interview with anonymous UPDF MI officer, Kampala, August 7, 2010.

379    Uganda group presentation, Kampala, Uganda, 24 June 2010.

380    See the diagram (chart 1) for details.

was achieved. After UPDF succeeded in its pressure on the LRA, the LRA relocated into South Sudan counties of Magwe and Torit which are bordering Uganda. The LRA used these counties as their bases to launch its into Uganda. Thus, the LRA remained as a threat to Uganda and South Sudan. This necessitated another round of joint operations by the UPDF and the SPLA meant to flush out LRA fighters from these counties. This operation pushed LRA to Western Equatoria and again into the jungles of the DRC and the CAR. SPLA MI Brigadier Mac Paul. K describes:

> *This operation managed to flush out LRA in most parts of our borders with DRC though the LRA has continued to cause havoc in Western Equatoria State, the Iron Fist Operation, is one of the successfully operations conducted by the two armies of the sisterly states, namely southern Sudan and Uganda.*[381]

Throughout intelligence sharing exercise, funding by the governments (GoS, GoU, and GoSS) towards intelligence operations was very poor. It was revealed that neither the government of Uganda nor the government of South Sudan that had fully committed itself to meeting the cost of the intelligence gathering and analysis. Operations funding had been coming from the US and it was mainly for the support of UPDF. The SPLA was not considered in the US funding. This was despite the fact that (as it has been observed amongst South Sudanese intelligence officers interviewed), the prevalent view was that 95 percent of LRA atrocities after 2007 were shouldered by the South Sudanese, and to some degree by the DRC and CAR. Yet the whole

---

381    Brig. Mac P. Kuol statement during border security meetings 8 July 2010.

support against the LRA went 100 percent to Uganda and its forces. As it has been already mentioned, this is an argument for the US to urgently consider once the further operations are considered.

Yet another factor was related to the community engagement. Community engagement was lacking and thus willingness of communities to support UPDF and SPLA with essential information was also lacking. The anti-LRA allies should consider providing social services such as clean water, health and educational facilities and construct roads, to encourage movement of people and goods in order to win support of local population. Coordinating institutions were still another problem identified throughout the exercise. Coordinating institutions such as the ones developed by European Union were missing. These institutions serve as arenas of exchange of views between forces. Neither the UPDF nor the SPLA put emphasis on the importance of such institutions. This clearly demonstrates there has not been any planning involved from the beginning of the agreement.

In intelligence sharing, financial support always remained a critical factor in the success of intelligence collection and dissemination. During intelligence sharing, Uganda was the only beneficiary. The support in many aspects went to the UPDF instead of the support being provided and distributed fairly to all those involved in the LRA operations for a comprehensive strategy and better coordination. The US, through its military's African Command (Africom), should provide communications, logistical and intelligence support not only for the Uganda army/UPDF, but also for other involved forces such as the SPLA, and others who are currently taking part in the fight against the LRA. This is to ensure the success of a strategy to catch the LRA leadership.

US involvement in the fight against the LRA has been due to the fact that the LRA has been categorized, as terrorist's organization. Thus, it was thinkable that the UN mission in Darfur, southern Sudan, DRC, CAR who were not working together, should have started to work together by sharing necessary information regarding the LRA. Also, French forces based in Chad and CAR must have been involved because the LRA had been defined as a regional problem rather than being Ugandan problem alone. The terrorist attacks in Belgium and France have once again highlighted the contradiction between the seemingly free movement of terrorists across Europe and the lack of EU wide intelligence sharing. Due to their earlier criminal activities, most perpetrators of the attacks in both Paris and Brussels were known to the various security agencies in several EU member states. For instance, the Abdeslam brothers had run a café in Brussels that was notorious for drug peddling. In early 2015, Belgian police questioned them about a failed attempt to travel to Syria, but they were not detained. Soon after, Dutch police stopped them during a routine traffic check, fined them for carrying a small quantity of hashish and then released them because they were not listed in their national information system. Allegedly neither the French security agencies nor the EU coordinating agency, Europol, were informed of either of these incidents prior to the Paris terrorist attacks in November 2015.[382] Similar stories of information non-sharing have emerged in the aftermath of other major terrorist attacks in Europe since the Madrid bombings in 2004. In response, EU policymakers have repeatedly promised to im-

---

382    La Baume, M., & Paravicini, G. (2015). Europe's intelligence 'black hole'. Politico. eu, Updated 12 August. http://www.politico.eu/article/europes-intelligence-black-holeeuropol-fbi-cia-paris-counter-terrorism/.

prove intelligence sharing across Europe, and some have even floated the idea that Europol should be turned into a centralized EU criminal intelligence hub, akin to the US Federal Bureau of Investigation (FBI).[383]

Regardless of the fact that local people and lower level LRA members were not the target of the ICC, the broader political environment has created a situation that has complicated human intelligence operations. The exceedingly rigid stance of the ICC has polarized the local context and thus the action of intelligence gathering from human sources has been hindered. Many in the LRA fear being given to the ICC and thus are no longer willing to give information. Moreover, many local people fear being implicated and thus do not talk to SPLA and UPDF.

The Hague must design a way to reconcile justice and peace. The major factor contributing the collapse of Juba Peace talks was the warrant of arrest by the ICC against LRA leaders.

To date, there has been nearly no coordination among the regional armies on facilitating the return of LRA captives who return from the bush. Therefore, it is hard to get information from them. There has been an ad hoc and poorly coordinated effort to deal with returnees on a case-by-case basis. It is critical that the regional governments establish a central reception centre, working with international agencies, to ensure that returnees are given the basic support and provisions needed to begin what is often a lengthy and difficult reintegration process. Also, there must be much more emphasis on protection for children who have recently been demobilized. Since neither the Congolese nor

---

383    Zimmermann, D. (2006). The European Union and post-9/11 counterterrorism: A reappraisal. Studies in Conflict & Terrorism, 29(2), 123–45.

the Ugandan army/UPDF or the GoSS/SPLA is in any position to address these concerns, UNICEF ought to be playing a bigger role.

In order for intelligence sharing to be more effective and reliable, the collision or joint forces sharing intelligence on the LRA- namely the UPDF, SPLA, SAF and the US special forces, must rely on local population and LRA defectors. In fact, the best effective method of intelligence collection especially in remote areas like South Sudan and DRC is through human intelligence. The LRA pursuing forces -UPDF, SPLA, and the US Special forces must make local population happy so that they get more out of them. Also, they could get information if they set up additional reception centres in DRC and Southern Sudan to help facilitate the return of LRA members who are struggling to overcome the many logistical and security hurdles to escape. Moving reception centres closer to high LRA activity areas would at least help to reduce some of those barriers and help ensure that those in forced captivity do not get caught up in the battle to catch or kill LRA fighters and commanders.

## Conclusion

The significance of intelligence sharing by the US, Uganda, Sudan and South Sudan against the LRA disabled the LRA activities in the borders of Uganda and South Sudan. Equally it reduced the size and capabilities of the LRA. Th.e reduction of the LRA capabilities was marked by the closure of the African Union Rapid Response Task Force (AU RTF) facilities and its redeployment out of South Sudan. This came as a result of the 6th Ministerial Joint Coordination Mechanism meeting in Addis Ababa, Ethiopia during the JCM of the Regional Cooperation Initiative for the elimination of LRA that took place in Addis Ababa on 30 March 2017. In this meeting, the US announced that its special forces would cease operations on 26 April 2017. Similarly, the Uganda government announced that its UPDF will withdraw from RTF immediately after the withdrawal of the US special forces. The withdraw of the US from pursuit of the LRA was a major setback in the regional fight against the LRA. As the US announced its pullback, the Uganda government also announced its troops' withdrawal and the SPLA in South Sudan eventually had to abandon the fight. In fact, South Sudan pointed out clearly that it would not be able to host the Headquarters of the RTF without the support of US special forces and UPD.

# Chapter 11

⊙∞∞∞∞∞∞∞∞∞∞∞∞∞∞∞⊙

# The ICC and
# the Indictment of Joseph Kony

## Introduction

This chapter examines the role of the ICC in the Uganda's LRA conflict. Right from the start of the LRA war, President Museveni believed in a military solution to end the northern Uganda LRA war. This approach was typically grounded on the realist philosophies of Thucydides, Machiavelli, Hobbes, Hans Morgenthau and the likes. However, the failure of this approach to completely deal with the LRA war caused Museveni to explore other opportunities that may disadvantage the LRA. These opportunities manifested itself in the involvement of international community powerful institutions such as the ICC. The ICC had to intervene because in the course of the conflict, serious war crimes of international community concern were committed – particularly intentionally directing attacks (regular use of torture, mutilation, murder and abduction) by the LRA against the civilian population not taking direct part in hostilities, have been committed with complete impunity.[384]

## The ICC Intervention

In July 2005, the chief prosecutor of the International Criminal Court (ICC) formally began an investigation into crimes committed by the Lord's Resistance Army (LRA) in northern Uganda[385]. This was followed by issuing arrest warrants for five LRA commanders including Joseph Kony, on twelve counts of crimes against humanity and twenty-one other counts of war crimes[386]. Crimes against humanity comprise a number of acts systematically directed against civilian population. The Rome Statute Article 7(1) lists these acts to include: murder, extermination, enslavement, deportation, or forcible transfer of a population; imprisonment, or other severe deprivation of physical liberty in violation of fundamental rules of international law; torture, rape; sexual among others.[387] Further, the Rome Statute defines attack directed against any civilian population as a course of conduct involving multiple commissions of the mentioned acts against any

---

384    Ssenyonjo, M. (2005). Accountability of non-state actors in Uganda for war crimes and human rights violations: Between amnesty and the International Criminal Court. Journal of Conflict and Security Law, 10(3), 405-434.

385    Clark, J. N. (2010). The ICC, Uganda and the LRA: re-framing the debate. African Studies, 69(1), 141-160.

386    UN Integrated Regional Information Networks (IRIN), "Uganda: ICC Issues Arrest Warrants for LRA Leaders" October 7, 2005; available at www.irinnews.org/report.asp?ReportID=49420SelectRegion= East_AfricaSelectCountry=UGANDA.

387    International Criminal Court, Rome Statute of the International Criminal Court and related documents available at http://untreaty.un.org/cod/icc/index.html

such population pursuant to or in furtherance of a state or organizational policy to commit such attack.[388] Also, it defines war crimes in great detail and more importantly, the statute advances the definition of war crimes by extending it to offenses committed in internal armed conflicts; and it establishes jurisdiction over war crimes, in particular, when committed as part of a policy, or as part of a large-scale operation.[389]

The ICC intervened to answer the long-standing violence referred to them by the Ugandan government. However, as Kassaja Apuli writes that "The signing of the Final Peace Agreement (FPA) between the leaders of the Lords' Resistance Army (LRA) and the government of Uganda (GoU) was delayed, partly because the LRA leader Joseph Kony wants assurances concerning the (non)execution of the International Criminal Court (ICC) arrest warrants against him and other LRA senior officials[390]. This is in spite of the GoU assurance to Joseph Kony that once he signs the agreement, it will ask the ICC to defer the LRA case to the jurisdiction of Ugandan authorities"[391]. Thus, the ICC involvement was a complicating matter that hindered peace rather than bring peace[392]. Apuuli further writes:

---

388    Ibid.

389    Article 8 (1), of the Rome Statute. It draws heavily upon the definitions of war crimes arising from within the Geneva Conventions.

390    Apuuli, K. P. (2008). The ICC's possible deferral of the LRA case to Uganda. Journal of International Criminal Justice, 6(4), 801-813.

391    Ibid.

392    Opiyo Oloya, Careless Talk Rubs Salt in the Wound, NEW VISION (Uganda), Oct. 3, 2006, available at http://www.newvision.co.ug/PA/8/20/524630

*When peace talks opened in Juba in 2006, the warrants became a sticking point impeding the conclusion of the final peace agreement, with the LRA representatives arguing that the group would not conclude a peace agreement when the warrants were 'hanging around the necks' of their commanders.*[393]

Also, Prendergast and Rogoff echoed this point and argue:

*As a result, the ICC has become somewhat controversial in Northern Uganda. Having gotten no credit in providing the first significant point of leverage on the LRA in 20 years and helping to drive the LRA into a peace process, many inside and outside Uganda are now calling for a removal of the ICC from Uganda as it is perceived to be a principal impediment to a peace deal.*[394]

Similarly, Skye Wheeler has this to say:

*Twenty-two years of horrific war, and two painfully slow years of negotiations, setbacks, caveats, and compromises have come to this: the Ugandan government and the rebels of the Lord's Resistance Army (LRA) stand on the brink of peace, hindered by one last condition—the removal of the International Criminal Court's (ICC) arrest warrants. The LRA's leader, Joseph Kony, along with two other commanders, has refused to appear in Juba, Sudan to sign a final peace agreement,*

---

393    Apuuli, K. P. (2011). Peace over justice: The Acholi religious leaders peace initiative (ARLPI) vs. the International Criminal Court (ICC) in Northern Uganda. Studies in Ethnicity and Nationalism, 11(1), 116-129.

394    Prendergast, J., & Rogoff, L. (2008). R2P, the ICC, and Stopping Atrocities in the Real World. ENOUGH!

*citing fear of extradition to the ICC.*[395]

Despite, concerns related to the ICC involvement in the ending conflict through justice, which has become an impediment to the solution itself, nevertheless, the role played by the ICC has been remarkable. Joseph Kony is accused of being savage and being brutal to civilians in pursuit of his political objective. The crimes he has been accused of are murder, abduction, mutilation and the burning the property of civilians. These inhumane activities made Joseph Kony and other LRA field commanders infamous. Some of the acts that Joseph Kony and his LRA applied, such has mutilation and cutting off parts of human body, had not been done in Islamic organizations such as Al Qaida. These LRA unbelievable atrocities feature in the International Crisis Group report where it wrote:

> *By forcefully recruiting civilians to become porters, sex slaves or fighters, the original Acholi leaders are able to maintain numbers and continue their violent existence in the bush...But the impact on those who live in the region is devastating ... Since 2008, the LRA has killed more than 2,400 civilians, abducted more than 3,400 and caused an estimated 440,000 to flee their homes in fear*[396]

Perhaps, the LRA strategy of purposeful use of violence to achieve political ends was not what Kenneth Payne's definition of strategy meant. Payne defines strategy as the purposeful use of violence for

---

395    Skye Wheeler, "Uganda rebel Kony to sign peace deal in bush," Reuters, 29 March 2008, http://africa.reuters.com/top/news/usnBAN930636.html

396    International Crisis Group. The Lord's Resistance Army: End Game? Africa Report No. 182, 17 November 2011.p.i.

political ends[397]. However, this has not been the case. Any strategy that leads to human suffering is likely to be opposed by those who have humanity in their hearts. This explains why the Regional Cooperation Initiative for the Elimination of the Lord's Resistance Army (RCI-LRA), a regional mechanism of the AU authorized in 2011 and launched in 2012 in Juba was initiated.

The involvement of the ICC, however, has been met with strong criticism from both inside and outside Uganda. Firstly, some commentators insist that peace must come before justice and that the ICC's arrest warrants undermine prospects for peace in Uganda. For instance, in the Yale's Law Journal *Fish* argues that the Rome Statute should be amended to provide a greater opportunity for peace negotiations to succeed in situations where the ICC has indicted one party to those negotiations[398]. Indeed, the last 30 years, amnesties have been instrumental in halting human rights abuses and restoring peace in countries like Angola, Argentina, Brazil, Cambodia, Chile, El Salvador, Guatemala, Haiti, Honduras, Ivory Coast, Nicaragua, Peru, Sierra Leone, South Africa, Togo, and Uruguay.[399]

Secondly, there are those who maintain that the ICC's involvement rides roughshod over already existing traditional, indigenous forms of justice in Uganda. According to Acholi customs, when an offender declares that he or she has committed a wrong, the traditional conflict

---

397     Kenneth Payne (2018). Strategy, Evolution, And War. Georgetown University Press. p.1.

398     Fish, E. S. (2009). Peace through complementarity: solving the ex post problem in International Criminal Court prosecutions. Yale LJ, 119, 1703.

399     RC Slye 'The Legitimacy of Amnesties Under International Law and General Principles of Anglo-American Law: Is a Legitimate Amnesty Possible?' (2002) 43 Virginia Journal of International Law 191 – 197.

management system is triggered. The dispute resolution process iden-
tifies certain behaviours as "kir," or taboo. "These behaviours may
range from the criminal to the antisocial-- violent acts, disputes over
resources, and sexual misconduct - including behaviour that would
prevent the settlement of the dispute." Clans must then cleanse the
"kir" through rituals which help to reaffirm communal values. Many
argue that these traditional mechanisms in particular, represent important channels for reintegration and reconciliation which can and
should be widely adopted.

However, other questions with regards to traditional justice have
been raised. For example, it has been underlined that a big concern
with the Uganda Acholi traditional justice is about its credibility. This
position is held by the international human rights body. The international human rights organizations point out that Uganda's court
systems lack the credibility, procedures, and impartiality necessary to
adjudicate international war crimes. They argue that because Kony
and the other LRA leaders have been accused of murder, rape, recruiting child soldiers, and attacking civilians, no tenable peace is possible
without true justice.

As much as the ICC intervention in the Uganda LRA conflict was
intended to be a good thing meant to bring the suspects to trial and
be a deterrence to others who fight wars without limits, its timing,
especially when the conflict is in advanced processes of mediation, is a
subject of criticism. Flag bearers of this criticism were victims of LRA
atrocities themselves and those advocating on their behalf. Religious
leaders, in particular, the Catholic Archbishop, John Baptist Odama,
who heads the ARLPI, raised a big concern that the involvement of
the ICC would put an end to efforts to ensure a peaceful negotiation

of the conflict[400]. When the ICC released its warrants, Bigombe complained that the ICC had been hasty, as the ceasefire agreement she was close to brokering had been thwarted at the last minute[401]. Peter Onega, who headed Amnesty Commission in Uganda during LIRA crisis, noted that ICC warrants against LRA commanders frustrated his work by scaring away rebels who were thwarted at the last minute.[402]

The rationale behind international justice is the ability for the international community to punish criminals for committing atrocities. To many nations, this is perfectly reasonable, but for others, this sounds strange and threatening to their national legal system. The International Criminal Court (ICC) which is located in The Hague, Netherlands, has been active since 2002. Its purpose is to deal with "international crimes, where national legal institutions fall short". These crimes include genocides, crimes against humanity, and war crimes. For 18 years of its existence, the ICC has investigated, prosecuted and sentenced individuals who have been found responsible for the aforementioned felonies.

The idea of an international criminal court can be traced back to the end of the First World War. The League of Nations attempted to establish such a tribunal in 1937. Members of the organization failed to ratify it however, and the idea disappeared – until the end of the Second World War. Following the Nazis' genocide and crimes against humanity during the Holocaust, the idea was resurrected once again. The Nuremberg Trials, the first real attempt to create an international

---

400    W Pelser, 'Will ICC Prosecutions Threaten Ugandan Peace Process?' ACR, Issue 46 (28 Nov 2005); T Allen, 'Trial Justice: The International Criminal Court and the Lord's Resistance Army' (2006) 83 – 88.

401    See prudence, p. 132.

402    Ibid.

court, were run by the Allies – the United States, the United Kingdom, France and the Soviet Union – who acted as prosecutors, judges and jury.

In the early 1950s, further progress was made. The UN General Assembly created the International Law Commission (ILC), but in practice it lacked both influence and power. It took another forty years for anything more to happen. The early 1990s saw mass atrocities in former Yugoslavia and Rwanda, where special war crime tribunals were created to punish those responsible. In May 1998, efforts were made to establish an institution for international justice. During a diplomatic conference in Rome, the four categories of international crimes were established: genocide, war crimes, crimes against humanity, and crimes of aggression. The statute was adopted following a vote, with 120 countries in favour, seven against and 21 abstaining. Among the seven countries that voted against, only three have admitted to doing so: The People's Republic of China, Israel, and the US.

The Rome Statute entered into force on 1st July 2002, and the ICC was created following ratification of the statute by 60 countries. Ever since, the ICC has worked to investigate, prosecute and sentence individuals charged with crimes within the Court's remit. Investigations have officially been conducted in Uganda, the Democratic Republic of Congo, the Central African Republic, Sudan (Darfur), Libya, Côte d'Ivoire, Kenya, Mali and Georgia. Individuals such as Libya's former leader Muammar Gaddafi, presidents in Kenya and Côte d'Ivoire, Laurent Gbagbo and Uhuru Kenyatta, as well as the notorious Ugandan LRA Joseph Kony, have been indicted by the Court.

On the one hand, this reflects positively on the ICC; it has managed to conduct its work effectively. On the other hand, if one takes a closer look at who the ICC is recognized by, and where the Court can conduct its work, the outlook is different. Large and powerful countries such as the US and Russia have signed the Rome Statute, but have subsequently withdrawn their signatures, whereas countries such as China, India, Indonesia, Pakistan and Belarus have refused to sign it at all. The ICC cannot operate in countries that are not signatories to the Rome Statute, since its mission is not recognized. International institutions work only if they receive wide recognition. Regrettably, the world's most powerful countries do not recognize the ICC. Since the fear of being prosecuted for atrocities holds greater weight than pursuing international justice, the future is therefore dark. It is unlikely that this will change, but one can at least hope.

Although African Union member states such as Senegal, Niger, the Republic of the Congo and Uganda were instrumental in creating the Rome Statute, which is the treaty establishing the ICC, of late, tensions between the African Union and ICC have built up. At the heart of the disagreement has been a tangle of issues including immunity and procedural matters and the failure of the Court to broaden its membership. Although the tensions have subsided for now, the crisis of 2016 is symptomatic of strains in the international justice system. The challenge for the future will be how to build a fair system which is equipped to bring justice to the victims of conflicts in the world.

## Conclusion

Joseph Kony and his Lord's Resistance Army are famous for their wide violation of human rights which has gone beyond the Ugandan border. The LRA conflict is a regional issue directly affecting the lives of thousands of people in Democratic Republic of Congo (DRC), Central

African Republic (CAR) and South Sudan. Its systematic targeting of children, abduction, use of fear as a tool of war, and population displacement shocked the international community. The conflict's impact has also reverberated more widely, displacing people beyond the region and periodically drawing in external responses from the African Union, the United Nations, the International Criminal Court, the United States and the European Union, among other actors.

Perceptions about the LRA, its continued use of violence and prolonged lifespan vary across the region. In DRC, for example, many suspect the LRA presence to be a pretext for Uganda's exploitation of DRC's natural resources. In South Sudan, people see the LRA as an instrument of Khartoum, which they think is being used to destabilize their new country. They see the ultimate solution to the LRA as laying in Khartoum, not in the bush where the LRA operates. Within Uganda, the LRA is considered to be a symptom of poor governance. In CAR, the LRA presence is considered to be a spillover from another theatre of war and ultimately Uganda's problem to solve. Such perceptions point to a wider geo-political interest at play, the underlying feature of which is the long-standing antagonistic relations between Kampala and Khartoum, distrust between Kampala and Kinshasa, and a looming crisis between Sudan and South Sudan.

The Ugandan army has thus far succeeded in keeping the LRA out of northern Uganda, but in doing so, has not resolved conflict. Instead, it has displaced the violence deep into DRC, South Sudan and CAR. Statistics from UNOCHA estimate that as of December 2011, there were 465,696 civilians displaced.

The official military effort, led by Uganda's armed forces – with funding and support from the US since December 2008 – has been focused on containment of LRA actions, removing the LRA's leader Joseph Kony from the battlefield, and protecting families and communities from LRA attacks. Those who bear the brunt of the LRA's violent retaliations are therefore all too aware of the risks of a renewed military strategy. In research carried out in 2011 by peacebuilding organisation Conciliation Resources, conducted as part of an EU-funded project partnership with Saferworld, an overwhelming majority of those consulted expressed a desire for a solution based on protection and political engagement.

The US strategy to support the disarmament of the LRA, issued in November 2010, emphasises that "there is no purely military solution to the LRA threat and impact'". Between December 2008 and November 2010, the US provided more than $23 million to the Ugandan army for military operations largely in form of logistical (airlifts, fuel, tracks) and intelligence support. The figure has since risen to more than US$ 40 million. In October 2011, a hundred US military advisers were deployed in the region to advice militaries in the region. All of this military intervention is ongoing and yet the LRA attacks against civilians continue unabated. The LRA conflict requires a holistic and coordinated response.

In short, there's been no shortage of official attention on the chaos caused by Joseph Kony and his combatants, though seasoned commentators note that not all actors in this conflict have their heart truly set on its resolution. Into the mix stepped Invisible Children and its latest campaign tool. In bringing the conflict with the LRA to the attention of ordinary global citizens – particularly young people in the US and the industrialized world – the *Stop Kony* campaign has succeeded where two decades of campaigning work had so far failed. The campaign has undoubtedly brought the issue of LRA into the spotlight of the mainstream media. It might even bring pressure to bear on governments to protect their citizens and look after their welfare.

However, what's clear for now is that it has had an interesting knock-on effect. For the first time in a very long period, the Ugandan Government, opposition politicians, religious and traditional leaders and the communities in northern Uganda agree on one fact: The *Stop Kony* campaign does not reflect the realities on the ground and glosses over the complicated history of the conflict and search for solutions. Patrick Loum, the Coordinator of the Acholi Religious Leaders' Peace Initiative (ARLPI) noted that "there is nothing in the campaign that reflects Uganda of 2012". He asserts that contrary to the impact of the military escalation for which the campaign advocates, it was "the Juba peace talks [which] brought the peace enjoyed in northern Uganda and parts of South Sudan". This is where lessons from Uganda should come into play for the situation as it currently stands. It is time for a new strategy. In spite of being widely acknowledged that there can be no exclusively military solution to this deep-rooted conflict, the international community and governments in the region are not investing sufficiently in finding alternative solutions to the LRA problem. Experience from northern Uganda teaches us that a combination of successful peacebuilding by civil society, a legal framework provided by the Amnesty law, political dialogue (the Juba Peace Talks) and increased vigilance by the army was responsible for the current peace enjoyed in northern Uganda and parts of South Sudan (Central and Eastern Equatorial State). What have worked have been local initiatives. Finally, the ICC investigation has resulted in the indictment of the LRA's five most senior leaders. However, they are yet to be apprehended. Meanwhile, the issuing of the arrest warrants seems to have ended all hopes of resolving the conflict peacefully.

# Chapter 12

LRA Defeated but Still at Large

## Introduction

This Chapter analyses the current status of the LRA and makes an argument that the LRA disappearance from the scene following severe pressure induced on it by the Collison forces comprising of Uganda, US, and South Sudan does not rule out the coming back of the LRA in future, especially when opposition is amounting is South Sudan. Taliban in Afghanistan has just taken over the seat of government in Kabul after many years of disappearance from scene, forcing the president Ashraf Ghani to flee. Therefore, if Taliban can come back to power after twenty years of defeat, then, why not LRA.

## Joseph Kony is Not Dead

Writing in 2013, John Prendergast stated that: "Before GANGNAM, there was the viral Kony 2012 video, which made Lord's Resistance Army (LRA) leader Joseph Kony the world's best-known international war criminal overnight, but the man himself remains at large in the jungles of central Africa"[403]. Indeed, Joseph Kony as we speak today remains at large clearly suggesting the fight against him has failed. Also, it indicates intelligence failure given that the objective to capture or killed Joseph has not been achieved.

While the best-known failures in history[404] have been warning failure against surprise attack in Denmark and Norway in 1940; Pearl harbour and the Philippines in 1941; Russia in 1941; Korea in 1950; as well the Chinese attack on India in 1962 to mention just few; intelligence failure with regards to LRA has been the inability of the US, Uganda, Sudan and South Sudan to kill or apprehend Joseph Kony. Also, the intelligence has not been able to tell the world where exactly Joseph Kony is located. What is known though is that the LRA has been pushed away from the Uganda South Sudan border into DRC and CAR borders.

Joseph Kony's refusal to sign the South Sudan mediated Juba Peace Talk frustrated the mediators and the countries affected by the LRA violence, thus prompted joint military operations conducted by UPDF and SPLA against the LRA. The operation was also supported by the US intelligence staff and special forces. As the operations progressed

---

403    John Prendergast. Get Kony. Foreign Policy, No. 198 (JANUARY/FEBRU-ARY 2013), pp. 52-53 Published by: Slate Group, LLC Stable URL: https://www.jstor.org/stable/41726746 Accessed: 13-04-2020 14:20 UTC.

404    Michael Herman (1996). Intelligence Power in Peace and War. Cambridge University Press, p. 200.

with intensive military pressure, the LRA was left with no better choice but to look for better areas of concealment away from Uganda and South Sudan borders. Thus, it relocated into DRC and CAR, where it remained silent since then, but what does this silence mean?

This silence has been interpreted as the defeat for the LRA and success for the US, Uganda, South Sudan and AU. Despite that, however, questions remain from curious minds about the status of the LRA today. Is it dead and finished? Is it alive and still there? Answers for these questions may generate debate but most are likely to agree that the status of the LRA as we speak today is unclear and neither the Uganda government nor the US and other partners in intelligence sharing, can confidently declare that the LRA is dead or alive.

With this situation unclear, the future is left blurred for the people of northern Uganda particularly, as well for those who were terrorized by the LRA atrocities. This uncertainty compels the governments of Uganda and South Sudan that something has to be done about it - by Uganda and its allies in the LRA fight. History reveals that many rebel movements appear, disappear and re-appear after being out of seen for long time. For instance, several Eritrean liberation movements fought Ethiopian regimes for over three decades, only to rise again to defeat Col Mengistu government when the Ethiopian People's Revolutionary Democratic Front (EPRDF) and the Eritrean People's Liberation Front (EPLF) collaborated to fight it and were able to defeat Mengistu government in 1991[405]. As a result of this, Ethiopia became two countries following the referendum which brought Eritrea independence on 24 May 1993. Going by history, it is likely that the LRA will one

---

405    Samson Wassara. Why Conflict in South Sudan and Somalia is Beyond Prevention and Management. Africa Insight Vol 49(3), December 2019.p.102.

day resurface and fight back, as the case of Eritrea has shown. The LRA conflict in the north of Uganda emerged on the grounds of getting the rights of the marginalized Acholi people. As Baguma notes "The northern Uganda conflict that finally spread to the Democratic Republic of Congo (DRC), Central Africa, and Sudan originated in the marginalized Acholi ethnic group of northern Uganda"[406]. This view must be contextualized to understand the LRA's future.

After nearly a decade of split, and the effect of this split on the military operations against Khartoum government forces, the SPLA lost its liberated areas in Upper Nile, Bahr el Ghazal and in Equatoria and it was pushed and confined with a great deal of frustration into a small area along the Uganda South Sudan border and remained there virtually inactive till 1995, when it regained its strength upon reception of military support from Uganda, Ethiopia, and Eritrea. The successes of 1995 reactivated SPLM/A strength and forced the Sudan Government to go back to the negotiation table, which culminated into the signing of the CPA on 9 January 2005.

The CPA brought back stability to Sudan and opened the way for South Sudan to reciprocate by mediating peaceful settlement between Uganda and the LRA. The failure of the South Sudan mediated solution brought South Sudan into the conflict as it then joined AU forces to fight LRA. However, a power struggle within the ruling SPLM party brought the country back to civil war in December 2013, eventually dividing the SPLM into three factions namely; the SPLM

---

406    Baguma, Charles (2012) "When the Traditional Justice System is the Best Suited Approach to Conflict Management: The Acholi Mato Oput, Joseph Kony, and the Lord's Resistance Army (LRA) In Uganda," Journal of Global Initiatives: Policy, Pedagogy, Perspective: Vol. 7: No. 1, Article 3. Available at: http://digitalcommons. kennesaw.edu/jgi/vol7/iss1/3

in Government (SPLM IG), the SPLM in Opposition (SPLM IO), and SPLM Former Detainees SPLM FDs). This civil war annulled the fight against the LRA with US and Uganda as the main players had to withdraw from South Sudan.

**Future Inter-Rebels Alliance**

With the Revitalized Peace Agreement in place, it is not yet clear what policy be adopted on the LRA. Will the operations against it be revived? Does the US still have an interest in fighting LRA in Biden's era? If not will Uganda and AU go it alone? The Revitalized Agreement on the Resolution of the Conflict in South Sudan (R-ARCSS) which became effective this year, left out some rebel groups such as the National Salvation Front (NAS) of the former SPLA Deputy Chief of General Staff General Thomas Cirilo Swaka and South Sudan United Front (SSUF) of General Paul Malong Awan, the former SPLA Chief of General Staff. Before the rupture of peace on 7 July 2016, the SPLM/A IG, the SPLM/A IO and SPLM/A FDs were the main actors in the civil war which erupted in 2013. The July confrontation involving the government forces and the IO forces in Juba, changed the nature of conflict in the sense that it gave birth to new opposition forces - for example NAS and SSUF. This new development complicated security in the sense that it led to a proliferation of armed movements in the Equatoria and Bahr elGhazal regions[407]. As Wassara states "the number of rebel groups and local militia groups continued to rise, especially in Equatoria".[408]

Rebellion in Equatoria, such the NAS led by General Thomas

---

407    Ibid.

408    Wassara, p. 108.

Cirilo, is unique from other regional rebellions and is very significant to the LRA because it brings in new element of transnationalism, which can be easily exploited by the LRA. The NAS forces are drawn mostly from Equatoria tribes who are spread along the South Sudan and Uganda border. Tribes such as the Acholi, Kakwa, Madi, Loguara, and the Kuku are found on both sides of the two countries and are core force of the LRA.

People like Duale had argued that the notion of transnationalism can be utilized to give meaning to social processes and can be recipe for intractable conflict and civil wars in different scales and levels[409]. Drawing from his experience in the Somali Conflict, Duale argued that transnationalism took on a different form in Somalia, where armed movements were born inside the country in reaction to prolonged political dictatorship, only to find themselves promoting Somali irredentist ideology, which advocated for the creation of Greater Somalia State. This included the former Italian colony of Somalia, British Somaliland, Djibouti, the Northern Frontier District of Kenya, and the Ethiopian Ogaden region.[410]

Unless the hold out warring parties' group in South Sudan are brought into the government through peaceful politically negotiated settlement, chances of South Sudan Equatoria rebels forming alliance with LRA are high. I can emphasize here that South Sudan's borders can become free of conflict if the likelihood of these groups forming alliance with the LRA is avoided. Such alliances have ever existed between Uganda and Southern Sudan rebels. In 1986 when Museveni had just

---

409    Ibid, p.104.

410    Duale, A.Y., 2015. Less and More than the Sum of its Parts: The Failed Merger of Somaliland and Somalia and the Tragic Quest for 'Greater Somalia'.

taken control of Kampala and its government policy towards SPLM/A was not clear, Dr. John Garang, the Chairman of the SPLM/A, welcomed Peter Otai UNLA forces and hosted them in Boma where its officers went under training by the SPLM/A. Garang's intention was to form alliance with this group in the event that Museveni NRM/A became hostile to SPLM/A. As soon as Museveni offered a hand of support, the SPLM/A abandoned cooperation with this Otai group forcing them to surrender to Museveni. During the Sudan SPLM/A civil war, Southern Equatoria based militia groups for example, the Equatoria Defence Force (EDF) joined hands with the LRA to fight SPLM/A. It is reported that in the border area between Sudan's South Darfur State, the South Sudan Kafia Kingi enclave, which, Sudan still occupies in violations of the CPA, and CAR's Haute Kotto prefecture, LRA groups have succeeded in establishing regular transnational relationships with several actors[411]. Until this time of writing neither US, Uganda, South Sudan knows what are plans of the LRA. The LRA relocated to DRC, CAR and Sudan Western Darfur region to escape the amounting pressure from intelligence sharing against it by the US, Uganda, and South Sudan. Although Sudan is a partner in the arrangement its participation was minimal. It was never keen to abandon LRA for geopolitical reasons because it has never viewed Uganda and South Sudan as political allies, especially when South Sudan was heading towards a referendum, the outcome of which Sudan hoped would be unity for Sudan. Thus, Sudan's participation was a political game to please US and present itself as a forerunner in the fight against terrorist groups.

Indeed, Sudan's reluctance contributed to intelligence failure, thus,

---

411     LRA Crisis Tracker. The State of the LRA in 2016. March 2016. P. 8.

resulting into inability to capture of kill Joseph through intelligence sharing. Ideally countries share intelligence for positive reasons, not for evil purposes. They share intelligence to protect humanity and human life from brutal terrorist organizations lacking moral obligations such as LRA. It is to be recalled that the 9/11 attacks and subsequent sharply increased focus on counter- terrorism, have led most countries to expand the number and depth of relationships they have with foreign security and intelligence[412]. Despite Sudan's reluctance in participation, intelligence sharing continued resulting in the LRA being totally pushed far from Uganda. However, it remained hiding along the South Sudan borders with Sudan, CAR, and DRC. It is not clear why did the LRA chose to relocate across Sudan's border but it is assumed that they did so to continue maintaining ties with Sudan.

While of late, considerable attention has been paid to improve intelligence sharing among western countries in an effort to face common threats, Africa's countries continue to play dirty games between and among themselves. For instance, wars since independence have been wars of proxy; and the LRA war is one of them. These proxy wars are mainly leaders designed to keep themselves in power. Joseph Kony and its LRA survival is thus explained in thess terms. They have been hiding and their whereabouts are not known and neither US with its advance technology, or Uganda, and South Sudan, can confidently say where the LRA remnants and Joseph Kony himself are. With Joseph Kony his whereabouts not known, it is premature to declare LRA dead.

There are some rebel groups which continue to operate after their leaders are eliminated from the battlefield, however, the LRA is tied to

---

412     Richards, R. (2012). A Guide to National Security: threats, responses and strategies. Oxford: Oxford University press, p. 163.

Joseph Kony's personality and leadership, so much so that his demise or capture would most likely put an end to group activities[413]. After masterminding the attacks of September 11, 2001, Osama bin Laden managed to vanish for 10 years, until the US realized that its war with Al-Qaida, a scattered group of individuals who were almost impossible to track, demanded an innovative approach. By 2011, intelligence accurately pointed a compound where Bin Laden had been hiding, resulting in his death. Osama bin Laden was an elusive personality and his killing is one of the great intelligence achievements that the world has ever seen during our lifetime.

The original US, Uganda, Sudan and South Sudan plan behind intelligence sharing had basically focused on the capture and elimination of LRA leader Joseph Kony. This decision was reached in order to stop LRA havoc in Uganda and South Sudan, and later on DRC and CAR. This had prompted Uganda, South Sudan and the Democratic Republic of Congo (DRC) with US military technical assistance, to launch a joint operation against the LRA through a trilateral arrangement in December 2007[414]. As the operation continued, logistical and political challenges prompted the coalition to request AU support for a strong Regional Task Force (RTF) in 2011.[415]

Unfortunately, the US and the AU support towards LRA fight was interrupted by the eruption of civil war in South Sudan in 2013. This civil war in South Sudan became a blessing to the LRA and Joseph Kony himself as regional operations ceased against it -thus, giving them breathing space. It was expected that this lull (as a result of the

---

413     John Prendergast foreign policy, no. 198 (January/February 2013), pp. 52-53.
414     UNSC 27. 3. 2009.
415     PSC 22. 11. 2011.

South Sudan civil war) would

be used by the LRA to recruit, reorganize and relaunch military operations. This has not been the case, however, but the likelihood of the LRA resurfacing in the near or far future cannot be completely ruled out unless Joseph Kony and the remnants of the LRA are located, persuaded to accept peace with Uganda government or to be dealt with through other means.

## The Need to Revive LRA-Uganda Government Political Solution

If Taliban can come back after twenty years of defeat, then why not LRA.As a matter of emphasis, for total peace to prevail in the countries affected by the LRA war and avoid future atrocities of the LRA, the government of Uganda should reinitiate peaceful settlement of the conflict with the LRA no how matter weak the LRA is; so, that those LRA's combatants who are still in hiding in the forest of DRC, CAR and South Sudan are encouraged to abandon rebellion and accept normal lives in their own countries. This is very important step as those who are unlikely to return voluntarily out of fear of reprisal, can have an opportunity to return through formal means. They can only accept voluntary return when they receive some assurance of their safety from the government of Uganda.

Similar situations from other countries can be used as lessons. In this way it is possible for the US, Uganda, South Sudan, DRC and CAR to solve the LRA problem in an uncostly way and save the lives of soldiers of their countries who are hunting for the LRA in the bush.

As a general assessment of intelligence sharing among the 'unequals' against the LRA, it is argued that the 2002 intelligence sharing agreement which continued for a decade has failed to produce a situation where intelligence has been fully effective in the operations against LRA. It failed to secure accurate information which would have finished LRA completely because of poor coordination, human error, insufficient funding commitment, and lack of proper training and skill for personal who are engaged in intelligence information gathering and analysis.

Interviews conducted in Uganda with several UPDF officers including the CDF[416] of Uganda and his Chief[417] of Military intelligence, indicated satisfaction with contribution of intelligence sharing towards the LRA fight. According to these interviewees, they concurred through intelligence sharing, a specifica reduction of LRA forces from the initial strength of about 5000 to a current force about of about 300 only. This claim could not be verified with LRA as it was difficult to establish contact with them. Critics however, questioned how the UPDF arrived at the projected LRA manpower number, and suggested that it is likely that the numbers given as LRA current strength by the Uganda Army are meant to meet media needs and public consumption. However, a Comprehensive Review of African Conflicts and Regional Interventions 2016 reported that more than 26, 300 rebels from 29 different insurgent groups defected and received amnesty[418]. Among these as reported, 49.3 percent were LRA

---

416     The author was privileged to have face to face interview with Gen. Aronda Nyaikairima, the Chief of Defence Forces of the Republic of Uganda by then.

417     I am grateful to Brigadier James Mugira, Chief of Military Intelligence, Republic of Uganda, for the interview with him on Thursday 13 May 2010, Kampala, Uganda.

418     Megan M. Gleason (ed.), Annual Review of Global Peace Operations – 2012: A Project of the Centre on International Cooperation, Lynne Rienner Publishers, Boulder, London, 2012, p. 40.

combatants.[419]

Indeed, by 2002 whenthe intelligence sharing agreement came into force, LRA operations and activities in Uganda and South Sudan reduced, mainly in Uganda and the Eastern Equatoria State in Southern Sudan, but remained active in Western Equatoria, and the neighboring DRC and CAR. 10 years later, as a result of the joint operations by Uganda and South Sudan's forces, the LRA's activities completely ceased in Uganda and South Sudan but shifted into DRC and CAR, where they continued to exercise their usual business of raping, abduction, looting and poaching. This prompted Uganda, South Sudan and the Democratic Republic of Congo (DRC) to launch a military joint operation against the LRA through a trilateral arrangement in December 2008. However, logistical and political challenges prompted the coalition to request AU support for a strong Regional Task Force (RTF).[420]

While Uganda and South Sudan forces were able to drive off LRA from their borders in their efforts to end the LRA rebellion militarily, the LRA shifted its operations into DRC and CAR and in a tit for tat game, launched several attacks on civil population in DRC and Central African Republic (CAR) throughout 2007 onwards. By 2012, the LRA has become a security threat in those countries too. Unfortunately, the civil war that began in South Sudan in 2013 reduced concentration and focus on the LRA, eventually with US withdrawing its financial

---

419    Sylvester B. Maphosa. The Lord's Resistance Army: A Review of African Union Regional Efforts to Eliminate the Resistance in Central Africa in Festus B Aboagye Ed. The Comprehensive Review of the African Conflicts and Regional Interventions, p. 20.

420    Brubacher, M. The AU Task Forces: An African Response to Transnational Armed Groups J. of Modern African Studies, 55 (2) (2017), Cambridge University Press, pp.277-299.

and technical support. Uganda and South Sudan also abandoned the operation in the absence of financial support from the US as it would be very difficult to maintain large number of troops involved in the operation.

The US has played its role and it is time for Uganda, South Sudan, DRC and CAR to note that the LRA is their enemy, not an enemy to the US. Therefore, they must make a serious commitment to maintain their troops in order to make the operation most effective. The starting point of this, is ownership of the whole process. Ownership of the process would bring commitment of these countries to inject resources on the LRA war, thus solving the problem of external financing, which can be problematic when they pull out - as happened with the US. The departure of the US from the agreement crippled operations, resulting in the concerned operations falling apart. This suggests that it is never a good idea to rely on one financier because many influential factors can occur, including donor fatigue. In intelligence sharing against the LRA, the burden of the first phase of operations burdens especially the financial aspect was on the US. It is to be reiterated however, the fact that the United States was an architect behind intelligence sharing agreement on the LRA does not mean that the US should carry the entire burden.

Another major weakness in the US, Uganda, and South Sudan intelligence sharing agreement has been the lack of institutionalized bodies to facilitate this move. Unlike, in the case of European Union, where three institutions - namely the Berne Group, Europol and the European Union Military staff, were constituted to facilitate intelligence sharing between its members states, US, Sudan and Uganda did not consider the role to be played by such institutions. They only relied on workshops meetings. A lots of funds were diverted to support these workshop meetings which were eventually without any significant contribution to intelligence sharing. Institutions established by the European Union serve the useful function of creating technical mechanisms for diffusion of intelligence among national authorities.[421] This is what US, Uganda and South Sudan should have adopted if they were really serious about their intelligence sharing. Furthermore, there had been inadequate preparation in the whole process of intelligence sharing.

Vital factors like climate and terrain were totally ignored from the very beginning. Among the intelligence officers interviewed, the prevalent view was that the jungles of Southern Sudan and DRC restricted mobility of intelligence officers throughout, leading sometimes to the delay of missions and to their tracks being easily identified. Additionally, the language of communication with local communities was a serious barrier. These factors affected adversely the accuracy of information. From interviews with intelligence officers involved in the operation, many acknowledged there had been interruptions related to weather and climatic factors. The region the LRA has relocated to is equatorial with high long thick trees and grass and heavy rains, especially for half the year from June to December. It is well known that weather changes, especially rains, interfere with reception of signals, hence creating inaccuracy and eventually leading to intelligence failure. However, for hindrances related to harsh weather and difficult terrain, human and technical intelligence have been, on the whole effective in the pursuit, as the case of the LRA by the UPDF and the

---

421    James I, Walsh, Intelligence Sharing in the European Union: Institutions Are Not Enough, JCMS 2006, Volume 44, Number 3, pp. 625-43.

SPLA have shown. This would mean that, with communication and signal intelligence being more unreliable in such conditions, human intelligence in terms of intelligence accuracy replaces communication and signal.

What should be done, though, in order to finish the LRA insurgence requires the main actors - the anti-LRA collaborators namely the USA, the UPDF and the SPLA, learn from the experience of operations Iron Fist and Lightening Thunder. These operations, despite all progress made have failed to achieve their ultimate objective- capturing or killing of Joseph Kony and his top commanders. This failure results from a failure to secure accurate information on the intended target. Poor co-ordination, human error, inadequate intelligence support equipment, shortage of finance, inadequate training, withholding of information, political restrictions and so forth all contributed to the survival of Joseph Kony and his top commanders.

A key problem identified with intelligence sharing among the unequal's is the disregard of information. As revealed by one interviewee, at times, valuable information was disregarded, for example in the Imatong Hills, where the LRA had been surrounded but managed to break through. According to Brig. David Manyok Barac now Major General, then SPLA 16th Brigade commander fighting the LRA, his troops had intelligence of the exact LRA position and possible escape routes, but his UPDF counterpart failed to act. As a result, the rebels sneaked back into northern Uganda.[422] When asked about the collaboration with UPDF, the SPLA sector commander by then, Gen Oyay Deng Ajak, whose forces fought LRA alongside with UPDF, acknowledged without elaboration the challenges facing the two sides in their

---

422    Interview with Brig. David Manyok Barac, Juba, 5 May 2010.

desire to get rid of Joseph Kony[423]. On his part, Lt. Gen Bior Ajang Duot, who was at the time of intelligence sharing the SPLA Deputy Chief of Staff for Operations, noted technical issues.[424]

Moments of intelligence failure are not new in history. On January 31, 1968, during the Tet holiday in Vietnam, North Vietnam's communist forces stunned the United States by launching a massive, coordinated assault against South Vietnam. While the communist military gains proved fleeting, the Tet Offensive was arguably the most decisive battle of Vietnam. Americans grew disillusioned with the war, prompting US policymakers to shift gears and focus on reducing America's footprint in Vietnam. A government inquiry shortly after the Tet Offensive concluded that US and South Vietnamese military officers and intelligence analysts had failed to fully anticipate the "intensity, coordination, and timing of the enemy attack" — despite multiple warnings[425]. Navy librarian Glenn E. Helm notes that disregard for intelligence collection, language barriers, and a misunderstanding of the enemy's strategy, played particularly prominent roles in the intelligence debacle[426]. Still, James J. Wirtz points out in *The Tet Offensive: Intelligence Failure in the War* that the "Americans almost succeeded in anticipating their opponents' moves in time to avoid the military consequences of surprise."[427]

---

423    Interview with Gen Oyay Deng Ajak, Juba, 5 June 2010.

424    Interview with Lt. Gen. Bior Ajang Duot, Juba, 11 June 2010.

425    Wirtz, J.J.1994. The Tet Offensive: Intelligence Failure in War. Cornell University Press.

426    Ibid.

427    Ibid.

## Another Phase of War with LRA May be Necessary

As I already highlighted, in a call for another chance to revive peace talks with the LRA combatants who are still hiding in the bush in DRC, CAR and South Sudan, it is the willingness of the LRA leadership which will determine the peaceful settlement of the conflict. In a situation where the LRA leadership is not willing to accept a call for peace, a need to go back to hunt the LRA is inevitable. In order for the objective of defeating LRA to be realized, it is critical that the US, and Uganda work closely with the government of the Southern Sudan, and other neighbouring countries, to plan and execute a military operation aimed at apprehending the LRA leadership based in north-eastern DRC. The United States and others concerned should provide much greater support for such operation, immediately, from planning to possibly execution on the ground for direct action against the LRA. Real-time information on the whereabouts of the LRA leader, Joseph Kony, and his henchmen is absolutely critical so that the Ugandan and Southern Sudan armies, with strong logistical support can respond quickly and precisely. The United States and its allies must also use their diplomatic influence to ensure that the Ugandan army is given access throughout the DRC and CAR, and that the Kinshasa Congolese government and that of CAR provide their full support commitment throughout the offensive.

Finishing the LRA insurgency requires that the main actors championing the fight must take advantages and learn from past operations. Poor co-ordination, human error, inadequate intelligence support equipment, shortage of finance, inadequate training, holding of information, political restrictions and above all, ignoring and neglecting the local population's participation in information sharing and intelligence gathering and so forth all contributed to the survival of Joseph Kony and his top commanders. In the US States for example, the Intelligence Community and the military work hand-in-hand to keep the nation and the deployed troops safe through shared intelligence and information. Each branch of the military has its own intelligence element tied together with population, which are in disguise become part of the military and part of the Intelligence Community. Together, these military and civilian intelligence community elements collect strategic and tactical intelligence which supports military operations and planning, personnel security in war zones and elsewhere, and anti-terrorism efforts. In war zones and other high-threat areas abroad, military and intelligence community civilians may be even co-located to maximize intelligence effectiveness.

Past experiences have shown that there are number of situations where the military acted on intelligence collected by non-military intelligence community components, such was the case with the raid on Usama bin Laden's compound in Afghanistan in 2011. The CIA collected valuable information on Bin Laden's whereabouts which the military used to execute an operation that resulted in Bin Laden's death. Similar to foreign liaison relationships, contact between military intelligence officers and their foreign military counterparts (who do not necessarily collect information military sources) has been critical to their own personnel protection and overall national security at home. This has been effective particularly when conducting joint maneuvers on foreign soil and dealing with transnational threats, as in the case of the LRA. These relationships fill information gaps, increase operational awareness, and improve understanding of local attitudes and tensions, particularly in areas where cultural complexities and norms are not well-understood by US personnel operating outside their own country.

Additional weakness in intelligence sharing against the LRA has been that information sharing relationships between the US, Uganda, South Sudan and Sudan did not take many forms, as is supposed to be. This includes the absence of mechanisms to deal with information gathered. There are formal mechanisms established to analyze, use, and disseminate information and intelligence in a way that it can be most useful to individual entities; these mechanisms were ignored. The SPLA, as host of the forces that came to follow up in the Kony operations was most of the time cut out of the picture, and only consulted on combat operations. There was no coordinating mechanism where intelligence where shared and discussed, unlike, the US where the Homeland Security and Law Enforcement Partners Board informs senior Intelligence Community officials and key federal partners, on the information needs and challenges facing state and local customers and communities. Also, such mechanisms participate in the US Department of Justice's Criminal Intelligence Coordinating Council which is composed of federal, state, local, tribal, and territorial government partners who convene regularly to identify and prioritize critical issues which can be addressed through improved intelligence and information sharing. Furthermore, the FBI sponsored Joint Terrorism Task Forces have been created in order to widen the breadth of the FBI's surveillance. Thus, information sharing between intelligence bodies are facilitated through online information sharing tools such as the Law Enforcement Enterprise and Regional Information Sharing Systems, which has never been established in the case of intelligence sharing on the LRA.

The US personnel involved in the LRA pursuit, assumably with their vast knowledge and experience on how intelligence is gathered, should have benefited from their past experiences at home or abroad. These soldiers know very well that at times, other countries have access to information that they do not have; such as in other parts of the world where local governments, paramilitary organizations, or populations are hostile towards the United States. The US, in this regard, has worked with other countries to collect and share information on transnational issues that affect them and the world: such as terrorism, cybercrime, drug trafficking, and weapons proliferation. These relationships would have been mutually beneficial in the LRA fight. These relationships would have developed liaison partners' such as Uganda and South Sudan capabilities to better pursue mutual threats, at the same time serve to leverage partners and networks to fill US information gaps, thus, strengthening US influence with their partners.

Good collection and good judgment are the key ingredients to producing good effective intelligence. The LRA capabilities were underestimated especially in the area of movements and manoeuvre, thus failing the test of good collection and good judgement. Further, it was assumed based on hearsay that the LRA was brutal and lacked relationships with locals without knowing that there were a few among the large population who may have had good connections with the LRA. Another related failure of intelligence in the pursuit of the LRA, has been the lack of using skilled analysts. The US, Uganda, and South Sudan should have considered by using the number of analysts in various agencies, thus, involvement of large workforce of intelligence analysts would at least statistically have increased the chances of launching surprise attacks on the LRA.

The opportunity of surprise attacks on the LRA were missed because US, Uganda and South Sudan forces failed to recruit within the LRA members who would have accurate information on their organization. Also, while the US managed to force Sudan to join hands with Uganda and South Sudan to fight LRA, it failed to make sure that Sudan was giving correct information thus leading to defection by states in information sharing arrangements. As McGruddy notes:

> While defection by states in information sharing arrangements is a real concern, taking over the control of the process by the strongest state is not helpful in the long term for the development of mutual trust, and certainly not in the best interests of the weaker participating state. A more multilateral approach would require developing an agreed set of standards for education, intelligence collection and analysis and protocols for sharing and collaboration, and of course, a strong and effective oversight body.[428]

Given the inherently secretive character of secret intelligence, there is immediately a tension between the need to maintain the secret, on the one hand, and sharing the secret or operating in a more open and collaborative manner on the other[429]. A highly accurate information may only be obtainable by planting a human insider within the enemy organization, which was not the case. A highly accurate information was obscured because the LRA was using sophisticated security and deception measures to conceal their real intentions. On the other side

---

428    Janine McGruddy Multilateral Intelligence Collaboration and International Oversight Journal of Strategic Security Volume 6, Number 5 Volume 6, No. 3, Fall 2013, p. 214.

429    Ibid.

of the coin, from the US, given the inherently secretive character of secret intelligence, there was immediately a tension between the need to maintain the secret, on the one hand, and sharing the secret or operating in a more open and collaborative manner on the other.

Also, it is worth noting that previous attempts to address LRA conflict politically were neither successful or fruitful. For example, there was the 1993/94 peace dialogue led by Betty Bigombe, who was Minister of State in Office of the Prime Minister, Resident in Northern Uganda, a post created by President Museveni to address the rebellion in Acholi. This was perhaps the most significant initiative led by the Government of Uganda (GoU) to engage directly with LRA in political negotiations to end the hostilities, unfortunately, the initiative collapsed before a final peaceful reconciliation could be realized.

In the aftermath of Operation Lightning Thunder, a peaceful way forward to end the northern Uganda war was difficult to see after Kony's repeated failures to sign the final peace agreement, clearly making peace impossible to achieve, thus, making us believe that the search for a military solution to the LRA problem has not succeeded. Indeed, Ugandan military attacks launched over the long course of the war have not only failed to defeat the rebels, but have typically made things worse for vulnerable citizens left unprotected in the aftermath of such operations. The US experiences in Afghanistan and Iraq should have made it painfully obvious that engagement with foreign countries in the form of diplomatic support and assistance with military intelligence, hardware, training, and advice can adversely result if a foreign military operation is not based on good intelligence and a well-conceived plan which includes clear and concrete objectives and has an overwhelming likelihood of success, and is then conducted in a way that those objectives are effectively carried out and followed up on. It is hard to imagine that Operation Lightning Thunder met those tests.

## Conclusion

As a matter of conclusion, this book in consideration of the current situation, argues that the LRA is still at large and thus remains an invisible threat. Because of this fact, the LRA will continue to terrorize the region at least psychologically until that time when Joseph Kony will come out of the bush voluntarily or killed. The Enough Project has correctly analysed this situation. As it states: "With a new and largely impenetrable base in the Central African Republic, Kony and his forces pose an immediate threat to neigh boring southern Sudan, north-eastern Democratic Republic of Congo, and southeast ern Central African Republic"[430].

Indeed, Joseph Kony and his LRA may not be a serious threat to their traditional Northern Uganda currently because it is far from his current operational base and better defended than other targets in the sub-region, but civilians in neighbouring countries of South Sudan, DRC and CAR are vulnerable, and if the LRA is not neutralized well in advance of the solution for the Abyei protocol between South Sudan and Sudan, there is a real danger that the Sudanese government will, as it has done in the past, use the LRA as a proxy force to destabilize parts of South Sudan.

The book argues that the intervention of the US forces had intentions beyond fighting LRA. Indeed, the US representative for African

---

430    Julia Spiegel and John Prendergast. A NEW PEACE STRATEGY FOR NORTHERN UGANDA AND THE LIRA. ENOUGH Strategy Paper 19 May 2008.

Affairs while complementing the efforts made by the African regional forces, suggested that these efforts would have impacts beyond the LRA. What the US envoy meant by "beyond the LRA" remains an unclear.

The LRA is seriously crippled by intelligence sharing led by the US but the sharing itself did not achieve the final goal to capture or kill Joseph Kony. The African Union's regional task force ramped up operations and exerted unprecedented pressure on LRA, resulting into fragmentation of its forces. Hence, the results of intelligence sharing were fruitful than, the previously unilateral military approaches that the Uganda Government had taken from the inception of the LRA. Operations such as Operation North, 1991, the Operation Iron Fist in 2002, and the Thunder Bolt were all part of the Uganda government's attempt to defeat LRA militarily.

# Bibliography

Acker, F.V, 2004. Uganda and the Lord's Resistance Army: The New Order No One Ordered. *African Affairs*, 103(412), pp.335-57.

Africa, S. and Kwadjo, J., 2009. Changing intelligence dynamics in Africa.

Agger, K, 2012. *The end of amnesty in Uganda: Implications for LRA defections*. Enough Project.

Ahere, J and Maina Grace. The never-ending pursuit of the Lord's Resistance Army: An analysis of the Regional Cooperative Initiative for the Elimination of the LRA. ACCORD, Issue 024, March 2013

Ajak, P.B, 2020. *My Escape to America Shows the Price of Dissent in South Sudan*. [Online] Available at: http://www.djreprints.com [Accessed 23 July 2020].

Aldrich, R.J, 2001. *The hidden hand: Britain, America and Cold War Secret Intelligence*. London: John Murray.

Aldrich, Richard J. (2009) Global Intelligence Co-operation versus Accountability: New Facets to an Old Problem, Intelligence and

National Security, 24:1, pp. 26-56.

Alexseev, M. (1997). *Without Warning: Threat Assessment and Intelligence.* St. Martin's Press.

Alexis, A; Lauren, P. The lord's resistance army: The US response. Current Politics and Economics of Africa; vol. 7, ISS. 2, (2014): 173-202

Allen, T, 2005. *War and Justice in Northern Uganda.* [Online] Available at: www.crisisstates.com/download/others/AllenICCReport. pdf [Accessed February 2005].

Allen, T, 2008. *Trial Justice: The Criminal Justice and the LRA.* London and New York: Zed Books.

Allen, Tim (2006), Trial Justice: The International Criminal Court and the Lord's Resistance Army, London: Zed books .... Article (1) Montevideo Convention 1933

Allison, G.T. and Zelikow, P., 1971. *Essence of decision: Explaining the Cuban missile crisis* (Vol. 327, No. 729.1). Boston: Little, Brown.

Amoah, M. (2011). Nationalism, Globalization, and Africa. USA: Palgrave Macmillan

Andrew, C., Aldrich, R.J., & Wark, W.K, 2009. *Secret Intelligence: A Reader.* Abington: Routledge.

Andrew, C. (2004). Intelligence, International Relations and 'Under-theorisation'. Intelligence and National Security, 19 (2), 170-184.

Anonymous. (2004). "How *Not* to Catch a Terrorist.» Pp. 50-52 *Atlantic Monthly* (December).

Anonymous. (2004). *Imperial Hubris.* Dulles, VA: Brassey's. Anonymous (Mike Scheuer). (2002). *Through Our Enemies' Eyes.* Dulles, VA: Brassey's.

Apuuli, K.P, 2008. The International Criminal Court and the Lord's Resistance Army Insurgency in Northern Uganda. In A. Nhema & P.T. Zeleza, eds. *The Resolution of African Conflicts: The Management of Conflict Resolution and Postconflict Reconstruction.* Oxford: James Currey.

Atkinson, R.R, 2009. From Uganda to the Congo and Beyond: Pursuing the Lord's Resistance Army. *New York: International Peace Institute*, pp.1-20.

Auten, J. (1985). "The Paramilitary Model of Police" In *The Ambivalent Force* **by A.** Andrew, Christopher, Richard J. Aldrich and Wesley K. Wark (2009), Secret Intelligence: A reader, London and New York, Routledge, p. 140.

Baguma, C, 2012. When the Traditional Justice System is the Best Suited Approach to Conflict Management: The Acholi Mato Oput, Joseph Kony, and the Lord's Resistance Army (LRA) in Uganda. *Journal of Global Initiatives: Policy, Pedagogy, Perspective*, 7(1), pp.31-43.

Banos, P. (2017). How They Rule The World: The 22 Secret Strategies of Global Power, London, Ebury press, p.4

Bassiouni, S.M, 1996. Searching for Peace and Achieving Justice: The Need for Accountability. *Law and Contemporary Problems*, 59(4), pp.9-28.

Bearden, Milt. The Main Enemy: The Inside Story of the CIA's Final Showdown with the KGB. New York: Random House, 2003.

Beesly, P. (2000). *Very Special Intelligence: The Story of the Admiralty's Operational Intelligence Center, 1939-1945*. London: Greenhill Books.

Behrend, H, 1991. Is Alice Lakwena a Witch? The Holy Spirit Movement and its fight against Evil in the North. In H.B. Hansen & M. Twadelle, eds. *Changing Uganda: The Dilemmas of Structural*

*Adjustment and Revolutionary Change.* Eds ed. Oxford: James Currey. pp.162-77.

Behrend, H, 1999. *Alice Lakwena and the Holy Spirits: War in Northern Uganda 1986-1997.* Oxford: James Currey.

Bellamy, C, 2002. *Governments 'Say Yes' as agreement is reached on global goals.* [Online] Available at: www.unicef.org/pressrelease [Accessed 25 May 2002].

Bennis, W. (1966). *Beyond Bureaucracy.* NY: McGraw-Hill.

Ben-Zvi, A. (1979). "The Study of Surprise Attacks" *Brit. J. of International Studies*, Vol. 5.

Betts, Richard. (1978). "Why Intelligence Failures are Inevitable" *World Politics*, Vol 31, No. 1.

Betts, R.K, 1978. Analysis, War, and Decision: Why intelligence failures are inevitable. *World Politics*, 31(1), pp.61-89.

Betts, R.K, 2007. *Enemies of Intelligence: Knowledge and Power in American National Sceurity.* New York: Columbia University Press.

Biersteker, T.J., Eckert, S.E., & Tourinho, M, 2016. *Targeted Sanctions: The Impacts and Effectiveness of United Nations Action.* Cambridge: Cambridge University Press.

Betts, R. (1982). *Surprise Attack: Lessons for Defense Planning.* Brookings Institute.

Betts, Richard K. (2007), *'Enemies of Intelligence: Knowledge and power'* in American National Security, New York, Columbia University Press, p. 264.

Behrend, Heike *'War in Northern Uganda'* in Clapham Christopher (1998) Ed., African Guerillas, pp. 107-118

Beren, F., *Can International Intelligence Sharing System Be Established For Global Security.*

Bevan, J (2007) The Myth of Madness: Cold Rationality and 'Resource' Plunder by the Lord's Resistance Army, Civil Wars

Booth, K, 2005. *Critical Security Studies and World Politics.* Colorado: Lynne Rienner Publishers.

Booth, K. Strategy and ethnocentricism. London, croom Helm, 1979

Born, Hans., Leigh, Ian. & Wills, Aidan. (2011) International Intelligence Cooperation and Accountability. Studies in intelligence series. London: Routledge.

Bullock, J., Haddow, G., Coppola, D., Ergin, E., Westerman, L. & Yeletaysi, S. (2005). *Introduction to Homeland Security.* Boston: Elsevier.

Blumberg and E. Niederhoffer (eds). NY: Holt, Rinehart & Winston.

Book, K, 1979. *Strategy and ethnocentricism.* London: Croom Helm.

Bramsen, I, 2019. Escalation dynamics of conflict displacement and violent repression in Bahrain and Syria. In I. Bramsen, P. Poder & O. Waever, eds. *Resolving International Cnflict:Dynamics of Escalation, Continuation and Transformation.* New York: Routledge.

Bramsen, I; Poder, P, & Waever, O, 2019. *Resolving International Conflict: Dynamic of Escalation, Continuation and Transformation.* Abington: Routledge.

Branch, A, 2007. Uganda's civil war and the politics of ICC intervention. *Ethics & International Affairs*, 21(2), pp.179-98.

Brubacher, M. The AU Task Forces: An African Response to Transnational Armed Groups J. of Modern African Studies, 55 (2) (2017), Cambridge University Press.

Bureau of African Affairs, 2018. *U.S relations with Uganda: BILATERAL RELATIONS FACT SHEET*. [Online] Available at: https://www.state.gov/u-s-relations-with-uganda/ [Accessed 28 October 2018].

Careau, F.H, 2004. *State Terrorism and the United States*. London and New York: Zed Books.

Carr, E. (1961). *What is History?* **NY: Vintage Books.** Chan, S. (1979). "The Intelligence of Stupidity: Understanding Failures" *Am. Pol. Sci. Rev.* 73:633-50.

Carter Center, 1999. *Agreement between the Governments of the Sudan and Uganda*. [Online] Available at: http://www.cartercenter. org/documents/nondatabase/nairobi%20agreement%201999.htm [Accessed 8 December 1999].

Chertoff, Michael. "Remarks by Homeland Security Secretary Michael Chertoff To The Heyman Fellows At Yale University On "Confronting The Threats To Our Homeland." 7 April 2008.

Clarke, R. (2004). *Against All Enemies*. NY: Free Press.

Clausewitz, Carl von (1976), On war, edited by M. Howard and P. Paret, Princeton, Princeton University Press.

Cline, L.E., 2016. African Regional Intelligence Cooperation: Problems and Prospects. International Journal of Intelligence and CounterIntelligence, 29(3), pp.447-469.

Clough, Chris. (2004) "Quid Pro Quo: The Challenges of International Strategic Intelligence Cooperation," International Journal of Intelligence and Counter Intelligence, 17:4, pp.601-613.

Congressional Research Service, 2019. *United States Foreign Intelligence Relationships: Background, Policy and Legal Authorities, Risks, Benefits*. [Online] Available at: https://crsreports.congress.gov/

R45720 [Accessed 15 May 2019].

Cox, R. *'Social forces, states and world order.'* Millennium: Journal of international Studies, 1981, Vol. 10. No. 2.

Christian Aid, 2004. *Uganda: Background to the Crisis.* [Online] Available at: http://www.christianaid.org.uk/uganda/background.htm [Accessed 20 February 2004].

Cox, R, 1981. 'Social forces, states and world order,' Millennium. *Journal of International Studies*, 10(2), pp.126-55.

Dahl, E. J. (2005). Warning of Terror: Explaining the Failure of Intelligence against Terrorism. Journal of Strategic Studies , 28 (1), 31-55.

Davies 1, P. H. (2004). Intelligence culture and intelligence failure in Britain and the United States. Cambridge Review of International Affairs, 17(3), 495-520.

De Temmerman, E, 2001. *Aboke Girls, Children Abducted in Northern Uganda.* Kampala: Fountain Publishers.

Dor, M.A, 2019. *Learning Through Negotiation: The Role of the SPLM/A in Ending Sudan's Second Civil War.* Perth: Africa World Books.

Dorn, W.A, 2009. Intelligence-led Peacekeeping: The United Nations Stabilisation Mission in Haiti 2006. *Intelligence and National Security*, 24(6), pp.805-35.

Duale, A.Y, 2015. Less and More than the Sum of its Parts: The Failed Merger of Somaliland and Somalia and the Tragic Quest for 'Greater Somalia'. In *Self-Determination and Seccesion in Africa*. New York: Routledge. pp.104-18.

Duke, Simon. (2006) "Intelligence, security and information flows in CFSP," Intelligence and National Security, 21:4, pp.604-630.

Dunn, K.C.2004. Uganda: The Lord's Resistance Army. Review of African Political Economy, 31(99),

Dupont, A. (2003). Intelligence for the Twenty-First Century. Intelligence and National Security , 18 (4), 15-39.

Durkheim, E. and Suicide, A., 1952. *A study in sociology*. London: Routledge & K. Paul.

Edwards, A, 2016. *Strategy in War and Peace: A Critical Introduction*. Edinburgh: Edinburgh University Press.

England, J, 2003. *Uganda Conflict Worse than Iraq: Interview with BBC News Channel*. [Online] The British Broadcasting Corporation [Accessed 10 November 2003].

Evans, G; & Newnham, J, 1998. *Dictionary of International Relations*. Harmonsworth: Penguin Books.

Fabricius, P, 2016. *The AU must grasp the nettle before Joseph Kony's Lord's Resistance Army returns to full strength*. [Online] Available at: www.issafrica.org [Accessed 6 July 2016].

Feldman, R.L., 2008. Why Uganda has failed to defeat the Lord's Resistance Army. *Defence & Security Analysis, 24*(1), pp.45-52.

Fish, E. S. (2009). Peace through complementarity: solving the ex post problem in International Criminal Court prosecutions. *Yale LJ, 119*, 1703.

Finnstrom, S, 2003. *Living with Bad Surroundings: War and Existential Uncertainity in Acholiland, Northern Uganda*. Uppsala-Sweden: Uppsala University Press.

Ford, Harold. (1993). *Estimative Intelligence. (2nd ed.)* **Univ. Press of America.**

Fromemer, L, 2002. *Uganda and Sudan join hands to fight LRA*. [Online] Available at: http://www.newsafrica.org/article/art [Accessed

May 2002].

Fry, M. G., & Hochstein, M. (1994). Epistemic communities: Intelligence Studies and International Relations. In W. K.

Fukuyama, Francis (2018). Identity: Contemporary Identity and the Struggle for Recognition. London: Profile Books.

Gaddis, J. (2005). *Surprise, Security & The American Experience.* Cambridge, MA: Harvard Univ. Press.

Geoffrey, R, 2002. *Crimes Against Humanity: The Struggle for Global Justice.* Harmondsworth: Penguin Books.

Ghani, A. and Clare Lockhart (2008). Fixing Failed States. Oxford, Oxford University Press

Gill, Peter (1994) Policing Politics: Security Intelligence and the democratic States, Routledge.

Gill, Peter. (2004). Securing the Globe: Intelligence and the Post-9/11 Shift from 'Liddism' to 'Drainism'. Intelligence and National Security, 19 (3), 467-489.

Gill, Peter. (2006) "Not Just Joining the Dots But Crossing the Borders and Bridging the Voids: Constructing Security Networks after 11 September 2001," Oxford Brookes University. Policing and Society. 16:1, pp.27-49.

Gill, P., & Phythian, M. (2006). Intelligence in an Insecure World. Cambridge: Polity Press.

Goodman, M.S., 2015. The Foundations of Anglo-American Intelligence Sharing. Laos: Operation MILLPOND, 1961 Foundations of Anglo-American Intelligence Sharing The National Intelligence Council, 2009–2014 Evaluating Insider Risk–The Critical-Path Method, 59(2), p.13.

Grabo, Cynthia. (2002). *Anticipating Surprise: Analysis for Strategic*

*Warning*. DIA-Joint Military Intelligence College.

Green, Mathew (2008), The Wizard of the Nile: The Hunt for Africa's Most Wanted, London: Portobello books

Gray, C.S, 2007. *War, Peace and International Relations: An Introduction to Strategic History*. New York: Routledge.

Green, M, 2009. *The Wizard of the Nile: the hunt for Africa's most wanted*. London: Gardner's Books.

Quaranto, P. J. (2006). Ending the real nightmares of northern Uganda. *Peace Review: A Journal of Social Justice, 18*(1), 137-144

Hatlebrekke, K. A., & Smith, M. L. (2010). Towards a new theory of intelligence failure? The impact of cognitive closure and discourse failure. Intelligence and national security, 25(2), 147-182.

Hart, G. (2003). "Post-Cold War Lassitude Contributed to the Attack on America." in M. Williams (ed.) *The Terrorist Attack on America: Current Controversies*. San Diego: Greenhaven.

Hart, P. (1990). *Groupthink in Government*. Baltimore: John Hopkins Univ. Press.

Herman, M. (2004). Ethics and Intelligence after September 11. in L. Scott, & P. D. Jackson (Eds.), Understanding Intelligence in the Twenty-First Century: Journeys and Shadows. London: Routledge.

Herman, M. (2001). Intelligence Services in the Information Age: Theory and Practice. London: Franks Cass. Hughes, G. (2011). Intelligence in the Cold War. Intelligence and National Security, 26 (6), 755-758.

Hulnick, A. S. (2006). What's wrong with the Intelligence Cycle. Intelligence and National Security, 21 (6), 959-979. Keegan, J. (2003). Intelligence in War: Knowledge of the Enemy from Napoleon to Al-Qaeda. London: Hutchinson. Laqueur, W. (1985). A World of Secrets:

The Uses and Limits of Intelligence. New York: Basic Book, Inc.

Heike Behrend (1999), Alice Lakwena and the Holy Spirits: War in Northern Uganda 1986-1997, James Currey Ltd.

Herman, Michael (1996), Intelligence Power in Peace and War, Cambridge, Cambridge University Press.

Hitz, Frederick P. & Weiss, Brian J. (2004) "Helping the CIA and FBI Connect the Dots in the War on Terror," International Journal of Intelligence and Counter Intelligence, 17:1, pp.1-41.

Hughes-Wilson, J. (2004). *Military Intelligence Blunders and Cover-Ups*. NY: Carroll & Graf.

Hulnick, A. (1986). "The Intelligence Producer-Policy Consumer Linkage: A Theoretical Approach." *Intelligence and National Security* **1(2): 212-233.**

Hulnick, A. (1997). "Intelligence and Law Enforcement: The 'Spies Are Not Cops' Problem." *International Journal of Intelligence and Counterintelligence* **10(3): 269-286.**

Institute for Strategic Studies, National Defense University of People's Liberation Army (2015) "International Strategic Relations and China's National Security." Singapore: World Scientific.

ICC, 2005. *"Statement by the Chief Prosecutor on the Uganda Arrest Warrant," The Hague.* [Online] Available at: www.icc-cpi.int/library/organs/otp/speeches/LMO_20051014_English.pdf [Accessed 14 October 2005].

Idris, A, 2018. *Why the US must not ignore the struggle of South Sudan's Soul.* [Online] Available at: https://thehill.com/opinion/international/419422-why-th-us-must-not-ignore-the-struggle-for-south-sudans-soul [Accessed 12 May 2018].

International Criminal Court, 1998. *Rome Statute of the International*

*Criminal Court*. [Online] Available at: http://untreaty.un.org/cod/icc/index.html [Accessed March 2008].

International Crisis Group, 2011. *The Lord's Resistance Army: End Game?* African Report No. 182,17 November. International Crisis Group.

Jeffery, K, 2010. *MI6: The History of the Secret Intelligence Service 1909-1949*. London: Bloomsbury.

Johnson, D.H, 2003. *The Root Causes of the Sudan's Civil War*. Oxford: James Currey.

Johnson, D.H, 2016. *The Root Causes of Sudan's Civil Wars: Old Wars and New Wars*. Oxford: James Currey.

Johnson, H.F, 2011. *Waging Peace in Sudan: The Inside Story of the Negotiations that Ended Africa's Longest Civil War*. Sussex: Sussex Academic Press.

Johnson, H.F, 2016. *South Sudan: The Untold Story*. London: I.B Taurus.

Johnson, L.K, 1997. Intelligence. In *Encyclopedia of US Foreign Relations*. Oxford University Press. pp.365-73.

Jones, C., 2007. Intelligence reform: The logic of information sharing. Intelligence and National Security, 22(3), pp.384-401.

Joost van Puijenbroek and Nico Plooijer. How EnLightning is the Thunder? Study on the Lord's Resistance Army in the border region of DR Congo, Sudan and Uganda. IKV Pax Christi, February 2009.

Kahn, D. *'Clausewitz and intelligence'* in M. Handel (1986), ed. Clausewitz and Modern strategy, London, CASS.

Kam, E. (2004). *Surprise Attack: The Victim's Perspective*. Cambridge, MA: Harvard Univ. Press.

Kauppi, M. (2002). "Counterterrorism Analysis 101." *Defense*

*Intelligence Journal* **11(1): 39-40.**

Kent, S. (1965), Strategic Intelligence for American World Policy, Hamden, Archon books.

Laqueur, Walter. (1985). *A World of Secrets: Uses & Limits of Intelligence.* NY: Basic.

Kettle, D. (2004). *System Under Stress: Homeland Security and American Politics.* Washington DC: CQ Press.

Kameri-Mbote, P, 2009. *The Land Question in Kenya: Legal and Ethical Dimensions.* Nairobi: Strathmore University Press.

Keohane R. and Nye, J. (1979). Power and Interdependence. Boston Theory of International Politics. Reading: Mass respectively

Kihika, KS. Evaluating the Deterrent Effects of the International Criminal Court in Uganda. In Schense, J. and Carter, L. (2016). Two Steps Forward, One Step Back, Deterrent Effects of the International Criminal Tribunals. International Nuremberg Principles Academy. P. 204

Laqueur, W. (1985). A World of Secrets: The Uses and Limits of Intelligence. New York: Basic Book, Inc.

Lee, K.Y., 2000. *From third world to first: Singapore and the Asian economic boom.* New York: Harper Business.

Lefebvre, Stephane (2003) '*The difficulties and Dilemmas of International; Intelligence Cooperation*', International Journal of Intelligence and Counter Intelligence, 16: 4:,527-542.

Lowenthal, M. (2003). *Intelligence: From Secrets to Policy, 2e.* Washington D.C.: CQ Press.

Lefebvre, S, 2003. The Difficulties and Dilemmas of International Intelligence Cooperation. *International Journal of Intelligence and Counter Intelligence*, 16(4), pp.527-42.

Lefebvre, S. (2003). The Difficulties and Dilemmas of International Intelligence Cooperation. International Journal of Intelligence and Counter-Intelligence, 16 (4), 527-542.

Lefèvre, M., 2010. Local Defence in Afghanistan. *Afghan Analysts Network*.

Leonard, E. *The Lord's Resistance Army an African Terrorist Group?* Perspectives on Terrorism, Vol. 4, No. 6 (December 2010), pp. 20-30 Published by: Terrorism Research Initiative.

Lowenthal, M. M. (2002). Intelligence: From Secrets to Policy. Washington DC: CQ Press. Odom, W. E. (2008). Intelligence Analysis. Intelligence and National Security, 23 (3), 316-332.

LeRiche, M; & Arnold, M, 2012. *South Sudan From Revolution to Independence*. London: Hurst & Co.

Levy, J.S., & Thompson, W.R, 2010. *Causes of War*. Malden-Oxford: Willey Blackwell

Lovelock, B, 2005. 'Securing a Viable Peace: Defeating Militant Extremists - Fourth Generation Peace Implementation'. In J. Covey, M.J. Dziedzic & L.R. Hawley, eds. *The Quest for Viable Peace: International Intervention and Strategies for Conflict Transformation*. Washington: US Institute of Peace Press. pp.139-40.

Lowenthal, M, 2003. *Intelligence: From Secrets to Policy*. 2nd ed. Washington DC: CQ Press.

Lefebvre, S. (2003). The Difficulties and Dilemmas of International Intelligence Cooperation. International Journal of Intelligence and Counter-Intelligence, 16 (4), 527-542.

Lowenthal, M. M. (2002). Intelligence: From Secrets to Policy. Washington DC: CQ Press. Odom, W. E. (2008). Intelligence Analysis. Intelligence and National Security, 23 (3), 316-332.

LRA Crisis Tracker, 2016. *The State of the LRA in 2016.* [Online] Available at: http://www.theresolve.org/wp-content/uploads/2016/03/The-state-of-the-LRA-2016-final.pdf [Accessed 21 January 2016].

Mamdani, M, 1976. *Politics and Class Formation in Uganda.* London: Heinemann.

Manjikian, M., 2015. But my hands are clean: The ethics of intelligence sharing and the problem of complicity. International Journal of Intelligence and Counterintelligence, 28(4), pp.692-709.

Maphosa, S.B, 2016. The Lord's Resistance Army: A Review of African Union Regional Efforts to Eliminate the Resistance in Central Africa. In F.B. Aboagye, ed. *A Comprehensive Review of African Conflicts and Regional Interventions.* Addis Ababa: African Conflicts and Regional Interventions. pp.212-64.

Marenin, O., 2006. Democratic oversight and border management: Principles, complexity and agency interests. *Borders and Security Governance. Vienna and Geneva: LIT Verlag/DCAF,* pp.17-40.

Marrin, S. (2003). "Homeland Security and the Analysis of Foreign Intelligence." *The Intelligencer* **13(2): 25-36.**

Marrin, S., 2004. Preventing intelligence failures by learning from the past. International Journal of Intelligence and CounterIntelligence, 17(4), pp.655-672

Martin, E.A. Ed. (2006). A Dictionary of Law. Sixth Edition. Oxford: Oxford University Press

McGruddy, J, 2013. Multilateral Intelligence Collaboration and International Oversight. *Journal of Strategic Security,* 6(3), pp.214-20.

Matei, F.C., 2009. The challenges of intelligence sharing in Romania. Intelligence and National Security, 24(4), pp.574-585.

Morganthau, H.J, 1991. *Politics Among Nations: The Struggle for*

*Power and Peace*. 6th ed. New Delhi: Kalyani Publishers.

Moynihan, D.P, 1998. *Secrecy: The American Experience*. New Haven: Yale University Press.

Museveni, Y.K, 2016. *Sowing the Mustard Seeds: The Struggle for Freedom and Democracy in Uganda*. 2nd ed. Nairobi: Moran EA Publishers.

Ngoga, Pascal 'Uganda: The National Resistance Army', in Clapham, Christopher (1998) Ed., African Guerrillas, pp. 91-106

Nugent, P, 2004. *Africa Since Independence*. London: Palgrave Macmillan.

Nussbau, M, 1997. Kant and Cosmopolitanism. In J. Bohman & M.L. Bachmann, eds. *Perpetual Peace: Essays on Kant's Cosmopolitan Ideal*. Cambridge/London: MIT Press.

Okoth, P.G, 2008. *Peace and Conflict Studies in A Global Context*. Ed ed. Masinde: Muliro University of Science and Technology Press.

Palan, R. and Blair, B. On the idealist Origins of the Realist theory of international relations. In Little, R. (1993). Eds. Review of International Studies. Volume 19, Number4 – October 1993, Cambridge University Press

Pateman, R. (2003). *Residual Uncertainty: Trying to Avoid Intelligence and Policy Mistakes in the Modern World*. Lanham, MD: Univ. Press. of America.

Payne, K, 2018. *Strategy, Evolution, And War*. Washington DC: Georgetown University Press.

Peace and Security Council, 2011. *Peace and Security 229th Meeting*. PSC 22. 11. 2011. Addis Ababa: African Union.

Pelser, W 'Will ICC Prosecutions Threaten Ugandan Peace Process?' ACR, Issue 46 (28 Nov 2005).

Pham, P; Vinck, P; & Stover, E, 2008. The Lord's Resistance Army and Forced Conscription in Northern Uganda. *Journal of Human Rights Quarterly*, 30(2), pp.404-11.

Prange, G., Goldstein, D. & Dillon, K. (1982). *At Dawn We Slept: The Untold Story of Pearl Harbor*. NY: Penguin.

Prendergast, J, 2013. *Get Kony. Foreign Policy, No 198*. [Online] Available at: https://www.jstor.org/stable/41726746 [Accessed 13 April 2020].

Prendergast, J, 2007. *What to Do about Joseph Kony*. [Online] ENOUGH Project: Strategy Paper No.8 Available at: https://www.americanprogress.org/wp-content/uploads/issues/2007/10/pdf/kony_report.pdf [Accessed October 2007].

Prunier, G. (2004). Rebel movements and proxy warfare: Uganda, Sudan and the Congo (1986–99). *African Affairs, 103*(412), 359-383.

Puijenbroek, J.V., & Plooijer, N, 2009. *How Enlightning is the Thunder? Study on the Lord's Resistance Army in the border region of DR Congo, Sudan and Uganda*. [Online] Available at: www.ikvpaxchristi.nl [Accessed 13 August 2020].

Rajapaksa, G, 2013. *Importance of Intelligence Sharing Among Nations*. [Online] Available at: http://www.asiantribune.com [Accessed 08 May 2013].

Randall, C, 2005. *Interaction Ritual Chains*. Princeton, New Jersey: Princeton University Press.

Ransom, H.H, 1970. *The Intelligence Establishment*. Cambridge, MA: Harvard University Press.

Reveron, D.S, 2006. Old allies, new friends: Intelligence-sharing in the war on terror. *Orbis*, 50(3), pp.453-68.

Richards, R. (2012). A Guide to National Security: threats,

responses and strategies. Oxford: Oxford University press.

Richelson, Jeffrey. The U.S. Intelligence Community. 5th Ed. Boulder: Westview Press, 2008

Schulsky, A. & Schmitt, G. (2002). *Silent Warfare: Understanding the World of Intelligence.* Washington DC: Brassey's.

Rose, C. and Ssekandi, F.M., 2007. The Pursuit of Transitional Justice and African Traditional Values: A Clash of Civilizations-The Case of Uganda. *SUR-Int'l J. on Hum Rts., 7,* p.101.

Rosenbach, E. & Peritz, A.J, 2009. *Intelligence and International Cooperation.* [Online] Harvard Kennedy School Available at: https://www.belfercenter.org/publication/confrontation-or-collaboration-congress-and-intelligence-community [Accessed July 2009].

Rudner, M, 2002. Contemporary Threats, Future Tasks: Candian Intelligence and the Challenges of Global Security. In N. Hillmer & M.A. Molot, eds. *Canada Among Nations 2002: A Fading Power.* Toronto: Oxford University Press. pp.141-71.

Russel, R. L. (2007). Achieving all-source fusion in the Intelligence Community. in L. K. Johnson (Ed.), Handbook of Intelligence Studies (pp. 189-199). London: Routledge.

Schelling, T. (1962). "Forward" in B. Wohlstetter, *Pearl Harbor: Warning & Decision.* Palo Alto, CA: Stanford Univ. Press.

Scott, L., & Jackson, P. (2004). Understanding Intelligence in the Twenty-First Century: Journey in Shadows. London: Routledge.

Salehyan, I, 2009. *Rebels Without Borders: Transnational Insurgencies in World Politics.* New York: Cornell University.

Schelling, T, 1962. Forward. In B. Wohlstetter, ed. *Pearl Harbour: Warning and Decision.* Palo Alto, CA: Stanford University Press.

Schmid, A, 2000. The Ultimate Threat: Terrorism and Weapons of

Mass Destruction. *Global Dialogue*, 2(4).

Schomersus, M., & Walmsley, E, 2007. The Lord's Resistance Army in Sudan: A History and Overview. *Gevena: Small Arms Survey.*

Schulsky, A., & Schmitt, G, 2002. *Silent Warfare: Understanding the World of Intelligence.* Washington DC: Brassey.

Scott, L, 2006. Secret Intelligence, Covert Action and Clandestine Diplomacy. *Intelligence and National Security*, 19(2), pp.322-41.

Seybolt, T.B., 2007. *Humanitarian military intervention: the conditions for success and failure.* SIPRI Publication.

Scott, L., & Jackson, P. (2004). Understanding Intelligence in the Twenty-First Century: Journey in Shadows. London: Routledge.

Seagle, A.N, 2015. Intelligence Sharing Practices Within NATO: An English School Perspective. *International Journal of Intelligence and CounterIntelligence*, 28(3), pp.557-77.

Seagle, A.N., 2015. Intelligence sharing practices within NATO: An english school perspective. International Journal of Intelligence and CounterIntelligence, 28(3), pp.557-577.

Seidman, H. (1998). *Politics, Position, and Power: The Dynamics of Federal Organizations, 5e.* NY: Oxford Univ. Press.

Sepper, E, 2010. Democracy, Human Rights, and Intelligence Sharing. *Texas International Law Journal*, 46(1).

Slye, RC 'The Legitimacy of Amnesties Under International Law and General Principles of Anglo-American Law: Is a Legitimate Amnesty Possible?' (2002) 43 *Virginia Journal of International* Law 191 – 197.

Smith, S. 'The Contested Concept of Security.' In Booth, K. (ed), Critical Security Studies and World Politics. Colorado, Lynne Rienner Publishers, 2005.

Stephen Lander. (2004) "International intelligence cooperation: an inside perspective", Cambridge Review of International Affairs, 17:3, pp.481-493.

Shpiro, S., 2001. The communication of mutual security: frameworks for European-Mediterranean intelligence sharing. Bar-Ilan University Department of Political Studies, pp.99-01.

Simonse, S., Verkoren, W., & Junne, G, 2010. NGO involvement in the Juba peace talks: the role and dilemmas of IKV PAx Christi. *The Lord's Resistance Army: Myth and Reality*, pp.223-42.

Sims, J.E, 2006. Foreign Intelligence Liaison: Devils, Deals, and Details. *International Journal of Intelligence and Counter Intelligence*, 19, pp.195-217.

Smith, S, 2005. The Contested Concept of Security. In K. Booth, ed. *Critical Security Studies and World Politics*. Colorado: Lynne Reinner Publishers. pp.27-61.

Southwick, K, 2005. *Investigating War in Northern Uganda: Dilemmas for the International Criminal Court*. [Online] Available at: www.yale.edu/yjia/articles/Vol_1_Iss_1_Summer2005/SouthwickFinal.pdf.

Sriram, C.L, Martin-Ortega, O & Herman, J, 2009. *War, Conflict and Human Rights: Theory and Practice*. Abingdon and New York: Routledge.

Staser McGill, A.K, & Gray, D.H, 2012. Challenges to International Counterterrorism Intelligence Sharing. *Global Security Studies*, 3(3), pp.76-85.

Strandquist, J., 2015. Local defence forces and counterinsurgency in Afghanistan: learning from the CIA's Village Defense Program in South Vietnam. *Small Wars & Insurgencies*, 26(1)

Sullivan, J.P, 2005. Terrorism Early Warning and Co-operation of Counterterrorism Intelligence. In *20th Anniversary International Conference*. Montreal, 2005. Canadian Association for Security and Intelligence Studies.

Twagiramungueta. "Re-describing Transnational Con ict in Africa." *Journal of Modern African Studies* 57/3 (2019): 378.

Temmerman, Else De (2001), Aboke Girls: Children Abducted in Northern Uganda, Kampala: Fountain LTD.

Tenet, George. At the Center of the Storm. New York: Harper Collins, 2007.

Todd, P., & Bloch, J. (2003). Global Intelligence: The World's Secret Services Today. Dhaka: University Press Ltd. Treverton, G. F. (2001). Intelligence and the "market state". Studies in Intelligence, 44 (2).

Uganda Resolve, 2009. *Uganda: As death toll in Central Africa reaches 1,000, Resolve Uganda calls for immediate action to protect civilians.* [Online] Available at: www.resolveuganda.org [Accessed 13 January 2009].

UN Integrated Regional Information Network, 2005. *Uganda: ICC Issues Arrest Warrants for LRA Leaders.* [Online] Available at: www.irin-news.org/report.asp?ReportID=49420 [Accessed 7 October 2005].

United Nations Security Council, 2009. *Twenty-Seventh report of the Secretary-General on the United Nations Organisation Mission in the Democratic Republic of Congo: S/2009/160.* [Online] United Nations Available at: www.monuc.unmission.org [Accessed 27 March 2009].

UNMIS, 2005. *Briefing note on the status of implementation of the Comprehensive Peace Agreement (CPA) in the Sudan.* Peace Keeping report 29 Sept 2005. UN Peace Keeping.

Viviers, W., Saayman, A. and Muller, M.L., 2005. Enhancing a competitive intelligence culture in South Africa. International Journal of Social Economics.

Walsh, James I. (17 May 2007), *Defection and Hierarchy in International Intelligence Sharing*, Journal of Public Policy, 27:151-181, Cambridge University Press.

Walsh, J.I, 2008. Intelligence Sharing for Counter-Insurgency. *Defense and Security Analysis*, 24(3), pp.281-301

Walsh, J.I, 2006. Intelligence Sharing in the European Union: Institutions Are Not Enough. *Journal of Common Market Studies*, 44(3), pp.625-43.

Walsh, J.I, 2009. *The International Politics of Intelligence Sharing*. New York: Columbia University Press.

Walter, L, 1985. *A World of Secrets: Uses and Limits of Intelligence*. New York: Basic.

Waltz Kenneth, N, 1959. *Man, the State, and War: A Theoretical Analysis*. New York: Columbia University.

Wark (Ed.), Espionage: Past, Present, Future? (pp. 14-29). lford: Frank Cass & Co.

Warner, M, 2002. Wanted: 'A Definition of Intelligence.' Studies in Intelligence. *Central Intelligence Agency*, 46(3).

Wassara, S, 2019. Why Conflict in South Sudan and Somalia is Beyond Prevention and Management. *Africa Institute of South Africa: Africa Insight*, 49(3), pp.102-13.

Whitlock, C., 2012. US expands secret intelligence operations in Africa. The Washington Post, 13.

Will Ross, Uganda army in 'rights abuses. BBC, Kampala, Uganda 16 July 2003.

Wirtz, J.J.1994. The Tet Offensive: Intelligence Failure in War.

Cornell University Press.

Wohlstetter, B. (1962). *Pearl Harbor: Warning and Decision*. Palo Alto, CA: Stanford Univ. Press.

Woodward, P. Uganda and southern Sudan 1986-9: new regimes and peripheral politics. In Wolfers, A. (1962). Discord and Collaboration: Essays on International Politics. Baltimore

Wright, Q, 1965. *A Study of War*. Chicago: Chicago University Press. V.1, 8.

Young, J, 2005. John Garang's legacy to the peace process, the SPLM/A and the South. *Review of African Political Economy*, 32(106), pp.535-48.

# Reports and Minutes

Crisis Group Africa Report N°157, 28 April 2010, 'LRA: A Regional Strategy Beyond Killing Kony' Gulu Communiqué Minutes, May 1, 2007

SPLA MI, Weekly Situation Report on incidence of insecurity resulting from tribal and intersectional fighting, LRA activities, SAF subversive activities, 4th -10th July 2009

SPLM MI Weekly Situation Report on incidence of insecurity resulting from tribal and intersectional fighting, LRA activities, SAF subversive activities, 1st -7th August 2009

SPLM MI Weekly Situation Report on incidence of insecurity resulting from tribal and intersectional fighting, LRA activities, SAF subversive activities, 8th -14th August 2009

SPLM MI Weekly Situation Report on incidence of insecurity resulting from tribal and intersectional fighting, LRA activities, SAF subversive activities, 15th -21st August 2009

SPLM MI Weekly Situation Report on incidence of insecurity resulting from tribal and intersectional fighting, LRA activities, SAF subversive activities, 21st -29th August 2009

SPLA, DRC, CAR & UPDF, The Report of the Joint Operations Review Meeting, 3rd – 4th September, 2009, Kampala, Uganda

SPLA MI, Joint Operations against LRA, 25 October 2009

SPLA Directorate of Military Operations, Operations Order No. 1, 11th December 2008, SPLA GHQs, Juba, South Sudan

SPLA MI, Weekly Situation Report on incidence of insecurity resulting from tribal and intersectional fighting, LRA activities, SAF subversive activities, 11th – 18th December 2009.

SPLA, SPLA brief notes on LRA Activities in South Sudan, Friday, February 19th 2010.

SPLA MI & UPDF MI, Surveillance Mission Joint Report, 25th March 2010: 1000 Hrs

UPDF MI Report, Brief to the SPLA COGS, 10th April 2010

The SPLA COGS and CDF of UPDF Meeting, 10th April 2010: Nzara, Yambio, Western Equatoria, South Sudan

# Newspapers

Anna Nimiriano Nunu, *An Interview with the LRA "Leader"* (2), Khartoum Monitor, Monday, May 24, 2010, Vol. 8 issue No. 1040, p.2

Anna Nimiriano Nunu, *An Interview with the LRA "Leader"* (3), Khartoum Monitor, Tuesday, May 25, 2010, Vol. 8 issue No. 1041, p.2

Patrick Okino '*Hunt for Kony to continue- Kiyonga*', The New Vision, Vol. 25 No. 96, Thursday May 13, 2010, p.8

'*American preacher sets out to find LRA's Koy,*' The Sunday Vision, Vol.17 No. 19, Sunday May 9, 2010, pp.18&19

# Interviews

## Sudan

General Oyay Deng Ajak, Minister for Regional Cooperation, GoSS, and former SPLA Chief of General Staff, Juba, Southern Sudan, 5 June 2010, Juba, Southern Sudan

Lt. General Bior Ajang Duot, Undersecretary, Ministry SPLA Affairs, GoSS, former SPLA Deputy Chief of General Staff, for Operations, Juba, Southern Sudan, 11 June 2010

Lt. General James Hoth Mai, SPLA Chief of General Staff, Juba, Southern Sudan, 11 June 2010

Major General Wilson Deng Kuoirot, SPLA Representative, 11 June 2010

Major General John Lat Zakaria, SPLA Director for Military Intelligence, 20 July 2010, Juba, Southern Sudan

Brigadier Malual Majok Chiengkuach, former SPLA Director for Military Intelligence, Juba, Southern Sudan, 29 June 2010, Nairobi, Kenya

Brigadier Mac Paul K, Awaar, Deputy Director, SPLA Military

Intelligence and Senior SPLA CJMC, 20 July 2010, Juba, Southern Sudan

Brigadier Riak Jeroboam Macuor, Military Intelligence, 26 May 2010, Juba, Southern Sudan

Brigadier David Manyok Barac, 5 May 2010, Juba, Southern Sudan

Col. Tut Jook, MI SPLA officer, 4 August 2010, Juba, Southern Sudan

**Uganda**

General Aronda Nyaikairima, Chief of Defence Forces, Republic of Uganda

Brigadier James Mugira, Chief of Military Intelligence, Republic of Uganda, Date of interview, Thursday 13 May 2010, Kampala, Uganda

Brigadier Charles Otema, UPDF Commanding Officer in Nzara, South Sudan, Friday 10 June 2010, Juba, Southern Sudan

Lt. Col. Richard Otto, UPDF MI Officer based in Juba, South Sudan, Saturday 7 August 2010, Kampala, Uganda

Lt. Col. Flex Kulayigye, UPDF Spokesman, Sunday 8 August 2010, Kampala, Uganda

# Index